EVOLUTION OF LEADERSHIP
— IN STEM —

WOMEN CATALYZING CHANGE

Curator's Email: cathy@inspiredtenacity.com
Curator's Website: https://inspiredtenacity.com
Email: lynda@actiontakerspublishing.com
Website: www.actiontakerspublishing.com

ISBN # (paperback) 978-1-956665-68-0
ISBN # (Kindle) 978-1-956665-69-7
Published by Action Takers Publishing™

Table of Contents

TABLE OF CONTENTS

Introduction

The STEM fields—science, technology, engineering, and mathematics—represent the cutting edge of innovation, the backbone of modern economies, and the keys to addressing humanity's most urgent challenges. From combating climate change to advancing healthcare and pioneering the technologies that define our future, STEM is where breakthroughs happen. Yet, for far too long, the transformative potential of these disciplines has been limited by a lack of diverse perspectives at the leadership table.

Evolution of Leadership in STEM: Women Catalyzing Change highlights the voices of women who have broken barriers, challenged the status quo, and redefined what leadership looks like in industries that were not historically built with them in mind. Within these pages, the resilience, creativity, and collaborative spirit that women contribute to STEM are showcased not just through their words, but through their impactful actions—demonstrating the remarkable difference they make when empowered to lead.

The Intersection of Barriers and Breakthroughs

The story of women in STEM is one of simultaneous progress and persistence. Although there have been significant strides in representation and recognition, systemic obstacles remain. Gender biases, unequal pay, lack of mentorship, and limited access to leadership roles are just a few of the challenges women face in STEM fields. These barriers are not just personal struggles; they represent missed opportunities for industries to benefit from the full breadth of talent and innovation that women bring.

In this book, these challenges are not shied away from—they are confronted head-on. Yet the narrative doesn't stop at the obstacles; it moves forward to explore the breakthroughs. Women in STEM leadership have not only overcome adversity but have also used it as a catalyst to build inclusive workplaces, foster collaboration, and champion a new era of leadership. Their stories demonstrate how change is possible—and how it is already happening.

Redefining Leadership in STEM

Leadership in STEM today looks different than it did a decade ago, and it will continue to evolve. The traditional models of top-down authority are being replaced by values-driven, collaborative, and inclusive approaches. Women leaders, in particular, are driving this transformation. They bring to the table unique perspectives shaped by resilience, empathy, and adaptability—qualities that are increasingly recognized as essential in tackling the complex, fast-paced challenges of the modern STEM landscape.

Through mentorship and allyship, women leaders are uplifting others, ensuring that the path they've carved is wider and smoother for the next generation. By embracing the power of personal branding, they are stepping into the spotlight, sharing their achievements, and dismantling stereotypes. And by cultivating emotional intelligence and leading with compassion, they are proving that effective leadership is as much about connection and collaboration as it is about technical expertise.

A Blueprint for Progress

The chapters in this book serve as a blueprint for what's possible when women lead in STEM. They explore key themes such as:

- **The power of mentorship and allyship** to foster diverse, equitable, and inclusive environments.

- **Strategies for breaking through systemic barriers** in male-dominated industries.

- **The importance of self-leadership and personal branding** in building confidence and visibility.

- **Reimagining workplace cultures** to support creativity, innovation, and well-being for all.

- **The role of emotional intelligence** in driving collaboration and building strong teams.

These are not abstract concepts—they are practical, actionable insights that readers can apply in their own lives and careers. Whether you're a woman navigating your path in STEM, an ally seeking to create more opportunities for diverse talent, or a leader committed to building an inclusive workplace, this book provides tools and inspiration to move forward.

Catalyzing Change Together

At its heart, this book is about the power of community. The evolution of leadership in STEM is not something any one person can achieve alone—it is the result of collective effort. Women leading in STEM are not just making history; they are shaping the future of these industries for the better. They are not only achieving success for themselves but are also paving the way for others, creating a ripple effect that touches lives far beyond their own.

This is a movement, and it is gaining momentum. The authors of this book represent the voices of many who are part of that movement—leaders, innovators, and changemakers who believe in the potential of a more equitable and inclusive STEM world. Their stories are both a reflection of how far we've come and a roadmap for how much further we can go.

As you turn the pages, let these stories challenge you, inspire you, and empower you. The evolution of leadership in STEM is here. Women are catalyzing change, and the future is brighter because of it. Together, let's champion this transformation and ensure that leadership in STEM becomes as diverse, dynamic, and innovative as the fields themselves.

Cathy Derksen – Curator

CHAPTER 1

Stepping into New Possibilities

Cathy Derksen

I dedicate this chapter to all women in STEM who are taking courageous action to bring about change that will empower and support the women around them.

"Brave leadership demands people to be courageous, self-aware, and put more importance on getting things right than being right" ~Brené Brown

---

My earliest memory of passion and curiosity was centred around biology. In my early preschool days, I had already developed a deep curiosity and respect for living things. I have always been curious about the magic of life and how it all comes together. As a young child, I collected and cared for all sorts of creatures, from dogs and cats to lizards, snakes, and caterpillars. I read books and watched documentaries about animals from every corner of the world. I was fascinated by the workings of my body and the magic that made it all work.

This passion has followed me throughout my life, so it was no surprise when I set out on a career that fed this interest. In university, I studied genetics and went even deeper with my understanding of the key components of life and living things. I love the fact that living things and our environment are so interconnected. Each depends on the others for survival in ways that we are still learning to fully understand.

After university, I was thrilled to find a career that allowed me to continue my passion for biology and lifelong learning. I specialized in clinical genetics and was fascinated to learn techniques to diagnose and explain a wide range of genetic variations and issues in humans. My work took me into hospital laboratories and public education.

In the early days, I saw myself spending a lifetime working in genetics. It was a field that was changing and expanding quickly with the field of genetics opening up into genomics and a whole new world of information to feed my curiosity and passion.

Unfortunately, this world lost its sparkle as I took a position under a doctor who believed that anyone who wasn't a doctor was not worth the energy to treat with basic respect or dignity. She treated her immediate team like ignorant nobodies and spread this negative behaviour out into the staff in the entire hospital. She would go so far as yelling at people for minor issues and give others the silent treatment if she wanted to make a point. It was worse than being back in high school.

She was in a position of power in the hospital and seemed untouchable, so there was nowhere to turn to bring about any change in this extremely negative situation. I knew that my mental health and my physical health were suffering every day I walked into work. After tolerating this working environment for several years and watching many of my talented co-workers pack their bags, I finally made the difficult decision to leave as well.

In a twisted way, this situation was a blessing. As I left behind a 25-year career, I was forced to look even deeper into my wide range of skills and interests in order to create my next career path. I rediscovered my love for working with people one on one. I recognized that I had a passion for helping people make positive changes in their life. The past 13 years have taken me on a journey of personal discovery and reinvention.

While I've stayed in touch with my previous community, I have also been blessed to participate in many international communities made up of women in STEM. I've met so many amazing women leaders who lead with a refreshing strategy incorporating collaboration, inclusion, and mentorship. Although we still have a long way to go to create a level playing field in regard to salaries and equal promotion of women and men in STEM fields, we have made significant progress in women's voices being heard and young girls discovering STEM as a career path for their futures.

My new career has taken me away from the hands-on work I used to do and I have left the world of STEM to step into new opportunities. My passion now is focused on creating books like this one to give women a platform for sharing their journey and their voice in a way that brings them together in community to create more impact. Stepping into my own style of leadership has allowed me to support other women on their journey. I believe that sharing our stories and wisdom is an impactful way of inspiring and motivating other women in our communities to take on new challenges.

Here are a few of the lessons in leadership I have learned on my own journey.

Self-leadership needs to be the foundation of all leadership possibilities. Learning how to manage yourself, your mindset, your emotional intelligence, and your social responses will establish your strength as a leader.

Lead with your own unique style. We all have skills to develop as we become leaders, but the lessons we learn should be focused on developing and enhancing our own style. When I first stepped into leadership roles, I was told I needed to be big and flashy. Whenever I attempted to develop that style, it felt inauthentic and uncomfortable. I learned that being myself and sharing my passion as authentically as possible is key.

Step out of the box and do things your way. Most of my life, I believed my career path needed to be made up of preexisting job descriptions. Now, I know that being creative in bringing your vision to life is so much more fun. The work I'm doing now, creating these multi-author book projects, was birthed from my passion of supporting women in sharing their stories and stepping into new possibilities in life. There were no guidelines to follow or job descriptions to stick with. My vision became my reality. Now, the work I'm doing impacts women around the world and shares voices that haven't been heard.

Build a community of support and encouragement. As you step into new areas of leadership, you will go through many challenges, and you'll need to navigate many obstacles along the way. It is critical to have a community around you that will provide a safe space to share your challenges, vent your frustrations, and remind you that you're not alone. Together, we can lift each other.

Be open to collaboration and mentorship. Find ways that you and your community can work together and provide complimentary services. You can achieve a new level of success by bringing together each person's expertise.

<u>Be courageous and bold</u>. Creating change in the world always comes with making major decisions and taking action in ways that aren't familiar or comfortable. Step into these challenges with confidence. Feel the fear and do it anyway.

<u>Dream big</u>. Allow yourself to create a big vision of the future you aim to create without the limiting beliefs that have kept you playing small.

<u>Be a role model for others</u>. As you step up into various levels of leadership in your life, other people will take notice. By stepping into a new version of yourself, you are giving other women the motivation and example to follow in your footsteps. Sharing your story and being open to supporting others creates a ripple effect of positive impacts on the world.

In my heart I will always be a woman of STEM and my love for biology and the magic of life will continue to be a part of who I am. I have found my own unique way of supporting women as they step into bigger possibilities in their life. Let's build a community of women focused on creating positive change around the world.

Cathy Derksen

Cathy Derksen is the founder of her company, Inspired Tenacity Global Solutions Inc. She is a Disruptor and Catalyst dedicated to improving the lives of the women in her community and around the world. Cathy helps women rediscover their brilliance, find their voice, and step into new possibilities in their life.

Cathy is an international speaker and 15x #1 bestselling author with stories that inspire the readers to take a leap of faith into reaching for their big goals. She has created a platform supporting women to share their own inspiring stories and wisdom in multi-author books including a wide range of themes. In her program, Cathy takes you from chapter concept to published bestselling author in a simple, exciting process. Cathy also creates retreat opportunities around the world for women to gather in person to build community and share their passions.

A decade ago, she transformed her career from working in Medical Genetics for 25 years to financial planning so that she could focus on helping women improve their life. After working as a financial planner for over a decade, she left her corporate job and started Inspired Tenacity to focus on helping women create success on their own terms and step into new possibilities.

Cathy has two children (31 and 30 yrs) and 2 fur-babies. She lives near Vancouver, Canada.

Cathy enjoys spending time in nature, traveling, meeting new people, and connecting with her global community.

Connect with Cathy at https://inspiredtenacity.com.

CHAPTER 2

One Aerospace Leader's Call for Change: A Reentry Story

Amanda Rose Fadely

This chapter is dedicated to my son Darion who didn't believe me when I insisted that I would never go back, and to the NG EM1 SLS Booster team for being there when I eventually did.

I consider myself fortunate to have had the opportunity to build a career in aerospace. The space industry has taken me on some incredible adventures and given me opportunities to be a part of history.

Yet, I did not follow a direct path to this life. In fact, if you'd known me during my high school years, you might have voted for me as 'Least Likely to end up as a Rocket Scientist'!

My parents valued education and were interested in a variety of topics including science, music, and literature. It wasn't uncommon for my mother's collection of pinned butterflies to cover the dining table or for my step-dad, an engineer, to set up a telescope for an evening of stargazing.

I had good friends and excelled academically. I was excited about the accelerated program that I was enrolled in for high school, as it promised students a competitive edge for college admissions.

But in the summer before high school, an experience at the hands of two older boys changed everything. It was a turning point in my life. Afraid or ashamed to tell my parents what had happened, I grappled with the resulting pain and confusion on my own.

Freshman year started, and I was a different person. Intellectual curiosity and drive were gone and I replaced them with drugs and parties. I managed passing grades during freshman year, but the turbulence in my life was escalating.

Sophomore year, I distanced myself further from both home and school. I found myself in increasingly dangerous situations and graduated to harder drugs. Eventually, my habits caught up with me, and I spent the last month of that school year in a drug rehabilitation center.

Thanks to either a teacher's pity or my mother's begging, I was granted a passing D grade in English and promoted to my junior year of high school. I moved back in with my parents and gave sobriety my best shot. Unfortunately, I was unsuccessful and things only got worse.

By spring, my parents realized that, despite their best efforts, they could not save me. They still knew nothing of the events that summer before high school. I was told that it was time for me to move out. At 16, I packed a single bag and left.

I moved in with a boyfriend several hours away. With even easier access to drugs, I became more reckless. But it wasn't until my boyfriend moved out of state with his parents and left me to stay with acquaintances, that I began to realize how much trouble I was in. I was hungry, broke, strung out, and now - alone.

Through a serendipitous series of events that is a story all of its own, I happened upon some money that gave me the means to visit my boyfriend. I finished up the last of the drugs that I had during two long, hot days of travel on a Greyhound bus.

He picked me up from the bus station and took me to his family's modest but comfortable home. I was grateful for at least a temporary safe place to stay and especially thankful for the kitchen full of food. A few weeks later (just before my 17th birthday), I would learn how truly providential my trip was. I discovered that I was nearly two months pregnant.

This news was the shock that I needed. Some kind of maternal instinct awakened within me, and I suddenly found the courage and drive to quit doing drugs. I got a part-time job and enrolled in an adult high school education program.

My son was born in March, in what would have been the spring of my senior year of high school. I decided that starting college quickly was more important than a high school diploma, so I got my GED that summer and entered my first semester of community college in the fall.

I still seemed to have an aptitude for math and science. Encouraged by that and inspired by memories of late nights studying the stars with my step-dad, I chose aerospace engineering as my major.

I enjoyed familiar academic success, but with the added stress of living on strained resources, raising a child, and keeping up with the demands of a rigorous academic program I had to work much harder for my achievements. I said no to most extracurricular activities, and was unable to nurture the types of lifelong college friendships that many around me developed. But I was proud, and thankful.

In 2003, I graduated with honors, summa cum laude, with my then five-year-old son at my side. Things had come full circle, in a way. The future was bright.

I'd love to say, 'and the rest was history,' but that certainly wasn't the end of the story.

As a single mom, I was driven by a sense of urgency and had taken the first job offered to me. I would not yet be working on the space projects that I dreamed about, but I had secured an interesting role working on the F-35 program.

My managers were caring and supportive and I was given opportunities to learn and to shine. I was aware that I was a minority as a female and, unsurprisingly, I didn't know any other engineers who could identify with the path I'd taken to get where I was – but the work environment was welcoming. I felt like a valued member of the team.

A few years later, I landed my first job in the space industry. I would be working on launch vehicles (rockets!) for a government contractor at Kennedy Space Center (KSC). I had so much to do and to learn, but it felt like a dream!

The challenges I had overcome in the past were not lost to me, and in those early days of the job I was overwhelmed with gratitude.

Then, reality hit. Hard.

Much like my previous role, one of my primary responsibilities was to monitor work in progress and provide direction where needed. This required a lot of my time be spent in-person at the work site which might be the launch tower, clean room, or other locations. The launch tower was the worst, and it didn't take long for the daily routine to become predictable.

My arrival would be marked by the team's whistling, yelling, and crude sexual comments. It was disorienting, at best. Every day felt like an experiment with the best way to handle their behavior. My own direct management was no help.

If the team felt like ignoring my direction to start on a work task, they made it clear that there was nothing I could do about it. If they felt like cornering me to ask never-ending inappropriate questions about my personal life, even my sex life, they would.

I tried to tell myself that they weren't bad guys, but only that they weren't used to having a woman around. But it took its toll on me.

It didn't help that I rarely saw other women. I almost never saw a woman on the launch tower, and the clean room wasn't much better. There were no women on my team. I was thrilled when I met a small group of female drafters but, understandably, they had separated themselves from everyone else and worked in a closed room far from the engineering team. We rarely crossed paths.

Feelings of alienation and depression began to set in. I tried building relationships with coworkers, but it felt like a one-sided effort. I would feel hopeful after a good one-on-one work discussion with a colleague, only to be taken completely off-guard by the condescending comments they would make as soon as other teammates joined the conversation.

Even basic needs became a source of stress. One worksite that my team frequented had a reputation for being particularly bad for women. The first time I was there, I went to find the restroom only to be told that the women's restroom had been converted to the men's 'crapper' and that there was no restroom for me to use. This might have been funny if even one person had stepped up to help me find my way, to feel included, but no one did.

I knew all about conquering difficult circumstances, and I thought of myself as strong and capable. Yet, I couldn't figure out how to fix this. I had overcome so much; it didn't seem possible that after all of that I would be forced to give up my dream. I wanted to find a solution.

I was afraid to speak up since surely it would only give them more reason to prove that I didn't belong there. But every day was a struggle, and things weren't improving. Eventually I did speak up, and learned a hard lesson.

I learned that companies will at times go to great lengths to protect themselves from any appearance of wrongdoing – regardless of the cost to the employee. The response from HR was that there had been no wrongdoing. They concluded that it was all just miscommunication, 'a misunderstanding.' Nothing changed. If anything, it made things worse.

I was experiencing textbook sexual harassment in an environment that had a reputation for its lack of diversity, lagging decades behind modern, progressive workplaces. Logically, it made sense. But in the echo chamber of my own mind, I struggled to reconcile my experience.

Being curious and a fast learner had become a liability instead of a strength. I was failing at the only job I would be accepted for - one where I should be quiet and make myself small, cute, and permissive towards men no matter how they treated me. My experience and capabilities were irrelevant and unwanted, and my value seemed to lie only in how much I would tolerate their advances.

Those boys from the summer before high school had used their strength and their numbers to take what they wanted from me sexually, casting aside the real me. This experience felt eerily similar and equally unacceptable.

But with no way to make things better, it took less than a year for my confidence and mental health to be fully shaken. I put in my resignation for the job that I had considered a dream come true. I was angry and hurt, but mostly, I felt a deep sense of personal failure and defeat.

The real tragedy about the nature of harassment, especially in the absence of a meaningful ally, is that shame or a sense of personal failure is often the default response.

I wish I could say I bounced right back, but my view of the space industry had been mangled and destroyed. I was grieving the loss of a hard-won dream and I struggled to untangle the experience in my mind.

I avoided the industry for years, keeping busy and satiating my hunger for learning in other ways. I homeschooled my son, and let our studies wander into space exploration. A seven-year-old daughter joined us by way of adoption, and I taught math, science, and aerospace technology courses part time.

But life has a way of bringing things back around.

My circumstances changed, and in 2014 I needed to return to work full time. By then, my kids were teenagers. I was not driven by desire, but by necessity. I managed to find a job but it was based at Kennedy Space Center, and I was nervous about what I was walking into.

Still, I couldn't help but notice that I also felt a tinge of vaguely familiar excitement.

Upon my return, I found that those historic buildings still held the ghosts of a culture that had hurt me deeply. The unwritten rules of conduct inside those gates versus outside were still there. As such, my teammates were not exempt from its influence. Still, they were different, and in time - I found that I was, too.

I'd been away from the industry for years, but life had taught me more than I'd realized. When I felt I was being treated unfairly, I stood my ground. And to their credit, my new colleagues responded with grace. They welcomed me into the fold and treated me with respect and kindness, as a valued member of the team.

Along with restarting my career in space, I gained a group of trusted friends, mentors, and lifetime supporters.

The world is such that I can never take for granted how far we still have to go in making the space industry truly inclusive. But thankfully, my passion returned full force.

Since then, I am fortunate to have led challenging projects within some of the most innovative and historically significant sectors of this decades' New Space Race.

My little family is all grown up. My daughter has everyone's dream job managing an ice cream shop. My son has a beautiful wife, two adorable children, and a career in aerospace of his own.

When I walked across that stage to accept my university diploma all those years ago, I thought I was done with the hard part. I never could have imagined the challenges that were still ahead!

I didn't set out to become a feminist or a voice for inclusivity, but I learned that experiencing discrimination or harassment firsthand has a way of igniting a passion for equity. I consider it both a responsibility and an honor to help pave the way for the brilliant young women and other underrepresented groups coming up behind me.

Men, remember that it isn't enough to just think of yourselves as 'nice guys.' Educate yourselves and never stop learning how to use your sphere of influence for good. Go the extra mile to make the space around you inclusive. You'd want someone to do it for your daughter, mother, sister, or wife, so do it for the women around you.

Ladies, we have enough challenges, so don't work against each other. Be one another's best allies; we are much stronger together.

Whether I'm busy managing an engineering program, or busy writing and speaking to champion for women and other underrepresented groups, I'm here and I'm not going anywhere.

I know I'm right where I belong.

Amanda Rose Fadely

Amanda Rose Fadely is an experienced space industry leader, real estate investor, entrepreneur, and author working on her first book. She loves the ocean and the mountains nearly as much as she loves the stars, and splits her time between Florida's Space Coast and the Upper Peninsula of Michigan.

Amanda's career in aerospace has provided her opportunities to work with incredibly talented international teams along with a collection of unique adventure stories to share with her grandbabies. Her work on both launch vehicles and satellites has taken her to Kennedy Space Center, Cape Canaveral Space Force Station, Vandenberg Space Force Base, Baikonur Cosmodrome in Kazakhstan, and Vostochny Cosmodrome in Russia. She has had the honor of supporting both NASA and commercial astronauts through training, launch management and spacecraft ocean recovery missions.

She is grateful for her family Nancy, Dennis, Scarlett, Darion, Kat, Max and Mira. They try to keep laughter and sarcasm to reasonable limits when they are all together, but nobody's perfect.

Learn more about Amanda's journey and explore her consulting, training, and guest speaking services and resources at www. amandarosefadely.com.

CHAPTER 3

Embrace Change and Say "Yes" to Navigate Your Unique Career Path of Success!

Amy Au

For my parents who always believed in me and gave me the gift of courage to follow my heart.

"A journey of a thousand miles begins with one step."
~Lao Tzu

When I saw the title of this book the first time, my gut feeling told me that I should contribute, but I hesitated as I questioned my ability to be an author. As I struggled to decide, the quote by Lao Tzu came to mind. If I did not take the first step, I would never know and I would not be able to achieve the goal of being an author.

Once I decided to take the step and look back on my career in STEM, I saw the impact of several key leadership skills I have had on my journey.

The journey from growing up in Hong Kong, to pursuing my education and career in Canada, has taken me through many challenges and opportunities for personal and professional development. My gut feeling is right.

I hope sharing these snapshots of my journey and how the leadership skills formed and supported my growth may resonate with you, encourage you if you are facing similar challenges, and inspire you if you want to walk the path in STEM.

Courage

"Courage nudges us outside our comfort zone where growth begins." ~Amy Au

---

The day has finally come! It was a bright, sunny September day. My first day of Grade 12 at a new school, a new city and country, in Vancouver, British Columbia, Canada. It felt surreal! The night before, I spent some time to pick out an outfit for my first day of school. In the previous 13 years, I didn't have to choose, school uniform was the norm in my life until then.

I didn't want to look overdressed. I didn't know about the school culture. Would I be bullied because of how I look and what I wore? Would I be laughed at? Would I be able to make new friends on the first day?

All these unknowns, questions filled my head.

I was a bit nervous, yet excited to be living on my own for the first time, thousands of miles away from my hometown, Hong Kong. English was not my first language, but I had learned as a language subject since Grade 1. "I should be fine," that's what I had been telling myself.

This is how my journey to leadership began.

When my parents asked me if I wanted to continue my education in Canada after secondary school, I didn't hesitate and said "Yes" instantly. The idea of being independent and responsible for my life from then on fascinated me. My parents' belief in me gave me the courage to take this big step in my life. I did not realize then how it would build my foundation of leadership.

Critical Thinking

"Critical thinking meets perseverance yields desired outcomes." ~Amy Au

---

Within the first year in Canada, I was 100% determined that I wanted to stay and live there. I was fortunate to have met great teachers who encouraged me to be the best I could be and made new friends who inspired me to choose the path of my own.

The question was, how would I be able to do that as a VISA student? It was unclear, but I knew I had to start planning in university early on. In the early '90s, professional immigration was encouraged, and the Technology sector was blooming. My interest had always been in science. It was natural for me to pursue a Science degree and I chose the University of Waterloo in Ontario, Canada. Since I graduated from High School in Vancouver, British Columbia, I took the leap to relocate to another province, Ontario, where the country's capital is and home of the biggest city, Toronto. It would be easier to find a job there after I graduated.

In my second year of University, I made a critical decision on my program. My major program was Science & Business, minor in Computer Science. My decision was based on three key reasons, 1) I could fulfill my passion on science, 2) I would pick up practical business skills, and 3) technology courses would allow me to find a job in tech and have the fast track on my professional immigration application. It was perfect!

Despite having extra courses with the minor program, I was able to complete my degree within four years as planned. The economic climate was not the best then; it was not possible for VISA student to get a job in the market after graduation. I took the swift decision to return to Hong Kong with the goal to have a job in the Technology sector and apply for immigration as soon as possible.

I returned to Canada within two years as a landed immigrant, which was within the shortest possible time. Mission accomplished, thanks to my critical thinking skill!

Positive and Continuous Learning Mindset

"Positivity is the key to unlock possibilities." ~Amy Au

My first job in Canada was a junior programmer in a software company. I was underpaid and I did not know how to speak up for myself then, but I needed a job to pay the bills and establish a new life, even though I felt mistreated in many situations. As the work environment turned toxic, it was time to move on if I wanted to grow. I decided to resign and looked for a new job.

Luckily, I found a new job within two months. I joined a global Fortune 500 company in the food manufacturing industry. It was a programming position in IT within a smaller business division, but I had many learning opportunities, and I grew in technical skills, soft skills, and business acumen.

One year in, an internal opportunity came up for me to join the Reporting and Analytics team in the corporate IT department. I learned then that my continuous learning mindset would be the key to my path of leadership development within the company.

In the next 10 plus years, I was the only female on the team, someone with the least knowledge on reporting analytics to picking up skills on database management, system administrative management, analytics tool design, project management, people management, stakeholder management, strategic planning and execution, and more. I developed from an analyst to manager, and then became the Director of Business Analytics.

I was proud of leading the team to execute a strategic decision on implementing a new visual analytics platform and began a transformative journey. It established a foundation of data analytics strategy and knowledge for future machine learning, predictive analytics solutions.

The success of my career progression was not without challenge. I was told to prove my ability repeatedly despite excellent performance reviews. There were times that I felt discouraged, disappointed, and imposter syndrome was very present at times. I did not know the term microaggressions to describe the situations I faced such as speaking up at meetings risked being ignored or dismissal of my opinion and ideas. It was challenging as a female in the male-dominated technology sector. I even delayed starting a family in fear of losing a promotion opportunity!

However, I had faced these challenges with a positive thinking mindset. I recognized the importance of building my mental strength from within. It made me stronger and started to learn how believing in myself fueled my energy to overcome the challenges and continue to learn and build soft skills that are critical as a leader. I was also fortunate to have managers and mentors who believed in me and supported my growth along the way.

During this time, both my positive thinking and continuous learning mindset helped me open new doors, explore new opportunities, and not be afraid to take on challenging tasks, managing difficult stakeholders' relationships. It made me an empathetic and inclusive leader.

Self-Confidence

"Self-Confidence energizes your inner superhero."
~Amy Au

In 2016, our local IT department was going to go through a few years of transformative change. Little did I know then that it would also be a pivotal moment for my career path. One day, my manager invited me to a one-on-one meeting and asked if I would be willing to take on a new role. The job description was vague, but I was compelled to say "Yes" despite uncertainty ahead.

I did not take the decision to leave my Business Analytics team lightly as I had a strong collaborative working relationship with every member on the team. However, it was necessary to fully focus on the transformation project and I intuitively knew that the opportunity would give me invaluable experience working directly with senior leadership in the regional and global levels. Also, I have developed much self-confidence over time with solid performance. I was grateful for the trust my manager had in me and the six years of adventure began.

Innovative thinking

"Make way for your innate creativity to shine in challenging times." ~Amy Au

---

In the six years of my IT transformation period, I observed the harsh reality of change impact: seasoned professionals who lost their jobs, attrition due to lack of confidence of the leadership, roles elimination due to redundancy, and others who accepted new roles for the paycheck. On the other hand, there were also positive impacts for some who embraced change, adapted quickly, and created new opportunities to advance their career.

I had the belief that there would be a path for me if I chose to keep an open mind, learn, and adapt. Instead of choosing to run away from the transformation, I chose to fully engage and create new pathways for me and support others in the team to embrace change.

During the six years of transformation, I took on three distinct roles. Each time gave me new opportunities to gain experience and grow. It expanded my horizon and developed further my leadership skills as a woman in tech leader.

I worked with global senior IT leaders to re-imagine and design the future of IT services from stand-alone local structure to a global, streamlined regional model to support local specific needs. The transformation was beyond a technical support change; it was a renewed approach to a business-driven focus. I learned much about organizational change management, organization design, talent management, stakeholders influence, and evolving teams to apply design thinking, and innovative thinking in both daily problem solving and new service and applications creation.

I became the regional North America team's IT Innovation Leader, a member in the IT North America leadership team. I led cross-functional Innovation Network champions to build a culture of Innovation, implement IT solutions that deliver business values, and transform the regional IT team during the challenging global pandemic period. It also marked my last role with this company after 23 years. There were fond memories, lasting friendships built through the years, and fruitful experiences that shaped my professional journey. While I was grateful to have a successful career in a global Fortune 500 company, I needed another pivot if I wanted to continue my growth.

As I closed one chapter of my journey in Technology Innovation, I opened another with my own transformative innovation!

Embrace Change

"Embracing Change broadens one's horizon." ~Amy Au

In Q1 2022, I had the most challenging time of loss in my life. My Dad unexpectedly passed away. I lost my job with a company that I worked for for 23 years. I had a choice then to either feel sorry for myself and give up, or to embrace change and look forward with a positive attitude. I chose the latter.

I had to admit it was overwhelming during the first six months, difficult to manage my emotion of losing a parent and losing a professional identity. I felt passive and not in control to navigate through the job-hunting routine. One day a light bulb moment hit. I needed to apply my innovative skills to re-imagine my future. I decided to take the unconventional path to start my entrepreneurial journey to become an Executive Leadership Coach.

I went through intensive training for six months, coaching with clients to gain my hours to fulfill the International Coaching Federation (ICF) accredited program as a professional coach. This step towards my new adventure was liberating and most fulfilling as I could directly create a positive impact for others.

I established my coaching practice and named my business "Innovate You." Anything is possible if you are willing to embrace change, try and learn from mistakes, and innovate yourself continuously. There is always another door of opportunity open. You just need to take the steps to walk through the door.

Resilience

"Resilience is a muscle; strengthening it benefits overall health." ~Amy Au

---

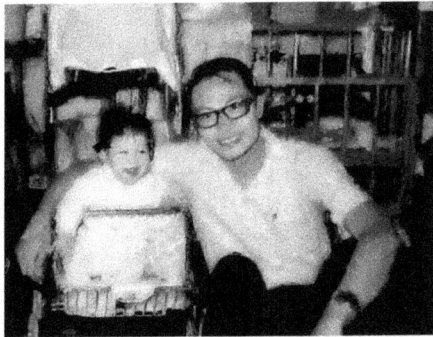

My dad taught me many things. His life motto, "Take things as they come," really stands out for me. He taught me that things do

happen with a reason and not necessarily within your plan. When that happens, face them with a composed state of mind. This advice helped me to overcome the tough times and re-focus my energy to build my new career as a coach. Over the years, this motto has had helped me build my resilience to face change in different situations, from the day I started a new life in a foreign country as a VISA student, to losing a long-time employment as a woman in tech leader. The "Take things as they come" and "Embracing change will bring new opportunities" mottos will continue to guide me through my life journey.

As a woman leader in tech, I have experienced firsthand the challenges and the much-needed support in the community to help other women leaders thrive. Therefore, I commit my coaching business to empower women leaders in STEM to reach beyond their perceived limits, evoke their innate creativity to innovate, unlock their potential, take confident actions to achieve their goals, and turn their aspirations to accomplishments.

My mission is to drive more positive impact and create a brighter future, one leader at a time. In turn, this will expand women in leadership, achieve diverse representation, reach gender balance and parity at senior management positions in STEM.

Working with my clients affirms the strength and experience I built over the years in the corporate setting. It prepared me well as an Executive Leadership Coach. I leverage the key leadership skills such as courage, critical thinking, positive and continuous learning mindset, self-confidence, innovative thinking, embracing change, and resilience to help myself and others to build a fulfilling life.

Although there are still many unknowns and challenges every step of the way, one thing I am certain is that I am on the right path to radiate inspiring transformation and my work is fully aligned with my purpose and values.

I hope sharing my story and my challenges can inspire you to pursue your dreams without hesitation. Take each step with confidence and be proud of yourself!

"Embrace change and say "Yes" to navigate your unique career path of success!" ~Amy Au

---

Amy Au

Amy Au is an ICF certified (ACC) Coach and Fascinate® Certified Advisor who radiates inspiring transformation. She founded her coaching business, Innovate You, in 2023 with the mission to drive more positive impact and create a brighter future, one leader at a time. She is passionate about empowering women in STEM leaders, professionals and teams to reach beyond their perceived limits, discover innate creativities and turn their aspirations to accomplishments faster.

Before expanding her career as an Executive Leadership Coach, Amy spent over 25 years in the technology sector, including 23 years at a global Fortune 500 company. During her last 10+ years there, she held people management and leadership roles in technology-driven innovation.

As a Fascinate® Certified Advisor, Amy uses the Fascinate® System to help her clients recognize their uniqueness and focus on what makes them most valuable, enabling them to present and communicate in ways that capture attention, build better relationships, stand out in the crowd, and build engaging and high-performance teams.

She is also an active certified #IAmRemarkable facilitator, volunteering to support women and underrepresented groups in practicing self-promotion. It helps build confidence, personal growth, and professional advancement.

Amy is based in Canada, and she partners with clients globally. She resides in Markham, Ontario, with her beloved husband, son, and a playful Siberian husky. She loves doing ballet, travelling, and hosting gatherings for family and friends.

Connect with Amy at www.InnovateYou.ca.

CHAPTER 4

Resilience and Reinvention: A Life in STEM Leadership

Amy Calder

To Joe, my unwavering rock. Your remarkable balance between work and life inspires me daily. Your strength, support, and ability to stay grounded through every challenge have been my greatest motivation. Thank you for believing in me and standing by my side.

At this point in my life, I have had some time to reflect on who I am, what is my purpose and who I can help. My journey is not one of straight lines or direct paths. I have meandered my way to my current life, and I wouldn't have it any other way. At this point I can divide my career path into four chapters of my life. I will share those chapters with you, but first I wanted to say a few things for you to get to know me better. I am about to turn 50 years old. For most of my life, I was the youngest person in the room as I climbed the career ladder, and now I am one of the oldest. I don't feel old, and I don't think old.

Recently I had the opportunity to re-take a personality assessment after I hadn't taken it for many years. It is the DISC profile, which many people are familiar with. When I first took the test, I was a "high-D." Almost off the chart in the Dominance category. I did a lot in my career to tone that down to work with others in the other categories. My recent result is on the line between Influence and Steadiness. I am no longer in the Dominance category, and I directly correlate that to my change of attitude prompted by a few medical scares. The extreme drive I had previously has been softened by my interest in forming and keeping strong relationships. Read on to learn more.

Chapter One: Early Aspirations

Starting out in my career, I was going to be a chiropractor or a veterinarian. As a result, I was studying biochemistry to prepare myself for either one. I should have studied more than I did, but I found a balance that allowed me to be social as well as get good enough grades.

I graduated from the University of Wisconsin-Whitewater with a major in Cell Biology and Physiology and a minor in Chemistry. I was ready for the next step. I still could not decide between Chiropractor and Veterinarian, so I got a job at a chiropractic clinic. It was not what I thought at all! It was a very repetitive job and all about volume of patients. The doctors were very open about having patients with good insurance coming back often and others left to decide when they needed treatment. Yuk! I didn't want to be part of the problem; I wanted to have impact and help people. Don't get me wrong, I appreciate the work of a chiropractor and I see one today, but I was not enchanted by the business part of that profession.

I moved on to work at a veterinary clinic. This was starting to feel more like me: helping animals and making a difference. I was working at the front desk when I took a call from an owner who wanted us to euthanize their animal because their fur was showing up on her new couch. Wow, do people really do this? That was the beginning of my awareness of how tough, and really thankless, the life of a veterinarian can be. There are really good days, but I saw a lot of bad days. Okay, check and check. I have ruled out the two things I wanted to be when I grew up.

Chapter Two: Finding My Way

I joined a temp agency and got a job at a local research company. I was staying in my field, but my interview consisted of me asking. "Do they bite?" I was going to be working with rats and mice in an animal research facility and it was scary! I took the job to be part of one of the largest local employers. I was a research tech running safety studies on rats and mice. It was not glamourous, it did not smell good, yes they do bite, and it was a very early start at 4:30am. I stuck with it for a few years. This quote comes to mind by Carlton Fisk, "It's not what you achieve, it's what you overcome. That's what defines your career." That not-so-fun job was the start of a long career—that is still going—in life sciences research. I stayed at that company for 21 years and it is where I really honed my leadership skills.

I became a supervisor and was leading a group of medical writers. I found my calling. However, I didn't realize it then (but I know now) that I used that role to prove myself to others. I was a tough manager. I did not hold back my thoughts and I said the hard things. Many times, without a thought about how I was received. Did I mention that at that time I was a high-D on the DISC profile? I wonder in hindsight if it was gender driven, but I didn't feel like I was working in a man's world. I just wanted to show my capabilities.

It worked for me. I was promoted almost every year of my employment there. It was during this time that I also obtained a master's degree in organizational management. I was given special opportunities and in fact moved from Wisconsin to San Diego and then to Austin pursuing opportunities with the same company. In Austin, Texas, as a 30-year-old, I was the youngest person to be the general manager of a Phase 1 Clinical Research Unit. This is where people go if they want to be paid to test new drugs. Pinch me! I was the leader of the entire site. My excitement didn't last long as the company decided to close several sites and mine was one of them. I was laid off.

Not to worry. Before my employment ended, another part of the company hired me to lead their client services team. I was thrilled and I moved back home to Wisconsin. Now I was part of the commercial leadership for a food testing division of our company.

I strongly believe I was given more opportunities because of the philosophy of the president of our company. We were a $2 Billion organization with over 100K employees, and the president was someone who understood utilizing skillsets. I was a leader first; that was my primary skillset. I loved leading teams and initiatives. I was not a technical expert. I did not retain my microbiology, chemistry, or toxicology knowledge. But I could speak the language and I felt close to the science. Our president recognized and rewarded good leadership. I was given special opportunities to grow my skills.

I took risks in my career. I had been leading the client services team in food testing and I accepted a special assignment to be a Six Sigma Black Belt leading commercial teams through change. I took the role knowing I risked being laid off. All the Black Belt roles before me had been eliminated over time, but I wanted the training and the experience of leading large change. As expected, I was again laid off when my role was eliminated. Much to my surprise, I was welcomed back to my previous role and served as the Global Client Services leader with staff in three countries and nine states for the next several years.

Throughout my career, I have taken the approach that the work will speak for itself. In other words, if I do good work, people will see, and I will be rewarded. This served me in the way that I wanted. After all, I avoided two layoffs. But I now see that maybe I could have been a better advocate for myself, specifically around salary.

Like many people who work in the sciences, I was then part of an acquisition. We were acquired by an even bigger company. Shortly after that, the larger organization sold my division to another company. I was no longer part of a large research company; I was part of a small food testing company.

I have since learned that leadership skills are less valued at less mature companies. Many companies in scientific areas feel strongly that you can't lead a team unless you are smarter at the science than your team members. This is unfortunate as it puts people into leadership roles for the wrong strengths, not because they can lead.

This leads me to Chapter Three.

Chapter Three: Navigating Personal Trials

Things were about to change for me, and I didn't know it but I was moving into Chapter Three. This chapter of my life has been shaped by health scares and a new attitude.

Shortly after my company was sold to a food testing company, my husband had a stroke (in his 30s, I might add). It was a shock. Thankfully it was not severe. He couldn't drive for quite a while and he now takes a daily dose of blood thinners, which is the only lasting outcome. However, during his night in the emergency room, the scan of his head showed something else. They told us to investigate it but wait until he was cleared of his stroke concerns. Okay, we did that. We saw an ENT a few months later to learn that he had thyroid cancer. He was cleared for surgery to remove it and we got that scheduled. While we were waiting for the surgery to happen, I was awakened in the night to my husband in so much pain he couldn't stand it and we went to the emergency room. Unrelated, but serious, his kidney was blocked and filling up with fluid. It was no longer functioning. He now needed a kidney removal after the thyroid surgery. (My husband's name is Joe, and we called that year his "Joverhaul.") That year we spent so much more time in hospitals than we ever want to again.

For the first time in my career, I asked for a break. I wanted to take three weeks off to care for my husband through his ordeal. I was granted the time, but my boss, one of those not-so-capable leaders that I spoke about previously, called me regularly for information and to do small tasks. I was not able to focus on my family like I wanted, and I was frustrated and decided to leave. Work is not everything and I needed to feel more supported.

I went to a new company in cancer research and found a very welcoming and supportive environment. It was great for a few years. I had many opportunities to work with amazing people.

However, chapter three is not over because that is when I learned of my own health issue. Two years after my husband's ordeal, I learned that I had Non-Hodgkin's Lymphoma. That was January 2020 and I started treatment in February 2020. (In case you don't remember, everything shut down in March 2020.) I didn't take a single day off. One of my big regrets is not taking a minute to rest even when I was at my most fatigued. It has left a lasting mark on my attitude about life and work. Thankfully, by the end of 2020, I was cancer free and healthy. I had a necklace made to commemorate my experience and to remind myself that Cancer did not define me or my life. It says, "a chapter not my life story." Life really IS short. Make the most of it.

This change in attitude was now solidified. I left my job. I was the chief of staff to the CFO at my organization. I was helping to arrange a few large layoffs and I decided I wanted to go, too. My boss agreed for me to be laid off with the next round and I took my severance package and left. Although it was a supportive company, their HR practices did not align with my personal philosophy. They hired a lot of senior people with little leadership skills and only technical expertise. During the time of layoffs, those leaders were protected and the frontline do-ers were regularly impacted. I didn't like being part of that, and I realized I didn't have to be any longer.

Chapter Four: Embracing Change

I have been on my own trying to make it work in my own consulting company for 2.5 years. Overall, it is amazing. I'm cancer free with all my hair and no lasting effects of the chemotherapy. There are some bad days. I didn't know until now how impactful imposter syndrome was going to be on my choices, but I'm moving in the right direction every day.

My personal business ownership journey has not been easy. I lost confidence in myself initially and purchased a consulting franchise. I thought this would be the net that would catch me, so I didn't fail. I would 'buy a job' and then build it bigger. After a year, I broke the contract. It was not a good fit, and it took me away from my passion to support the life sciences industry. It was a very expensive decision both emotionally and financially to walk away, but it was worth it.

I am passionate about helping others, change for good and making a difference. I am most satisfied when I am in a role surrounded by a lot of really smart people and when I am constantly learning. That is what has kept me tied to life sciences. I love learning what is next and what people are working on in their labs and how their technology can change the face of healthcare.

My company, Golden Innovation Group, offers life science companies the opportunity to outsource some of their mundane business operations tasks so they can focus on their science. The company name is intentional to pay homage to the beloved Golden Retriever whom I feel has qualities I want to emulate in my business such as respect, caring, and loyalty. We help companies set up streamlined workflows to get the most out of team members and not overspend. I have spent years of my career navigating personalities, politics, regulations and employees. I want to share that knowledge and help bring new inventions and developments to life.

These companies are critical to the fabric of our society, and I want to be there to help them. I join their team and we win together. I can't think of a more rewarding career as part of this chapter of my life.

In reflecting on my journey, I am reminded of the power of resilience, adaptability, and the importance of staying true to one's values. Each chapter of my career has brought its own set of challenges and triumphs, shaping me into the leader I am today. From my early aspirations to be a chiropractor or veterinarian to my unexpected detours into leadership roles and personal trials, every experience has contributed to my growth and understanding of what it means to lead with empathy and integrity.

I hope my story inspires other women in STEM to embrace their unique paths, to lead with courage and compassion, and to never underestimate the impact they can have on the world around them. Together, we can catalyze change and drive the future of science and technology forward.

Amy Calder

Amy Calder is the owner of Golden Innovation Group. She has a BS in Biochemistry and an MA in Organization Management. She has been an internal consultant to the life sciences industry for the past 25 years. She has executed large changes across global organizations. She's been involved in over 7 M&A deals on both sides of the table. In addition, she has held Operations, Commercial, and Compliance leadership roles.

Amy brings a unique perspective to the table for problem-solving given her diverse background. She has sat in many seats and understands the view from each of them. She is data-driven but also understands that the quest for perfect data can sometimes be a roadblock. Amy's network is vast and connects you to a variety of experts.

Amy believes that every company should operate precisely, efficiently, and purposefully. Golden Innovation Group's mission is to guide small to mid-sized organizations in transforming chaos into clarity. By streamlining workflows, documenting processes, and aligning teams, they are helping their clients by allowing them to focus on groundbreaking advancements while Golden handles the operational complexities.

Connect with Amy at www.goldeninnovationgroup.com.

CHAPTER 5

A Necessary Fall to Rise

Anu Ramchandran Nair

I dedicate my chapter to everyone who is trying to find a safe space for themselves. Never be afraid to be yourself. In the world of masks, let's bring some unalloyed and positive authenticity!

Since childhood, I've been captivated by science—the theories, formulas, and the intricate workings of the world around me. My unwavering fascination led me to pursue a path as a scientist. Why? Perhaps it was my curiosity with a pinch of societal expectations I grew in. Regardless, I've never regretted my academic journey. Every degree, every lesson—I cherish them all. And each day, whether in small ways or large, I contribute to making a difference.

If you are a good student, then you might have built an expectation for yourself, which is absolutely great giving yourself a goal to achieve; however, always remember the outside world is very different and exciting!!

I am excited to share my journey on emotional intelligence and how has it changed my virtue of life. Let me introduce myself. I am a science geek with immense love for creativity and innovation. I have never believed in selecting one particular lane and have embraced the fact that I am a multi-tasker. Is it difficult? Yes! It is, however, how we can embrace our uniqueness and the things that connect us with each other. I am someone who loves a good brainstorming! I knew I was a good leader, a good human; however, I did not like fights and learned a great way to avoid them. But, soon enough, I learned why it is important to voice yourself and why having healthy discussions (not arguments) are necessary. More than often I hear women called "emotional." It's so common that it is often mentioned comically all the time. I used to just laugh and ignore them until I was labeled as emotional. It made me furious. It wasn't funny anymore when I was put on the spot! I was frustrated and not only did I not know how to navigate my feelings, I didn't understand why I was labeled. My sense of dedication and loyalty were seen as "over-reacting," "attention-seeking," or "getting emotional." If you can relate, then this chapter is for you.

As a result of all the name calling, each time I would get overwhelmed, I would take deep breaths and isolate myself. My overwhelm became reactionary. Like a scholar, I started looking for the reason for this reaction. You know what it was? I wanted to be liked, so I did not confront.

For the longest time, I powered through, I changed my job, but I carried with me the feelings and reactions. I reached a point where I thought there was something wrong with me. I worked very hard towards my career and personal life. All I had ever known was if you need to reach somewhere, you need to work hard. One lesson I needed to learn was that, "No matter how hard you try, someone will always be unhappy with you!" You see, even though I knew this to be true, I still wanted to keep people around me happy.

Another lesson I learned was that, "You are not responsible for others' happiness and sorrow. You can help them and support them, but that is it!" I know, my dear reader, you must think, "Anu, how did you survive this world so far?," to be honest, it beats me!

Your first lesson:

Through everything I was going through, I never stopped standing up for myself. I fell again and then again. I fell so many times that I lost count. Those stories are for another time. Even when my personal life was suffering, I stood up for myself. I don't think it was because I was very strong; it was because that was the only option I had. I think that somehow we tend to attract what we want. So, I started from scratch again. I started networking from zero and made new contacts. This time, though, I started with a learned approach rather than an approach of just wanting to be liked.

Your second lesson:

During the ousting phase, I learned there is no point in blaming anyone. You are accountable for yourself, and I am accountable for myself. This phase led me to a journey of self-discovery. I reflected on my actions, and I started asking questions to myself no matter how uncomfortable it got. My dear friend, always remember we can never grow in our comfort zone; to evolve we need new perspective and to grow we need new space. I kept asking questions until I could find a solution to my problems. Like I said, it all starts within you. And yes, there was my answer: Boundaries!!!!

Today when I think back on my past, I realize I never set any boundaries. Even if I did, I was so scared to offend people and (being a people pleaser) I used to let things slide.

The toughest trait I had to ever work on was the need for acceptance. Is it nice to be accepted? Sure. But never make it a necessity. Once I realized how it is related to self-worth and confidence, it sent me on a path of growth. If you are on the journey of self-worth, let me tell you something, I have been there, too. I read books, reached out to so many people, and have watched so many videos to get to a point where I am comfortable writing it in a book and tell you! So, if you are on a similar journey, please don't give up. Listen to yourself, what is it that is blocking your energy? What actions can you take to overcome that? These were some of the questions I asked myself and they helped me.

Your third lesson:

When you are finishing school and entering the corporate world, doesn't matter how big or small a company is; always study the job environment you are in and remember a few things:

"Never be scared to voice your opinions."

"Be courageous." If you see yourself as a leader, you need to learn how to stand up for yourself. Think about it like this, if you don't stand up for yourself how will your team have any faith in you? How will you build yourself as a leader?

Also, you should know where to be silent. Woah! Is that a frown I see on your face? Are you confused? I said speak up, now I am saying not to? Naah! I am saying that a big part of being a leader is to know when to speak. Whom to speak to? And what to speak? Don't worry, these skills are easily learnt; I know you will be a great leader!

I had some rough experiences. Did they bother me at the time? Did I take a setback? Yes. Above all, though, did I learn? Yes, I did.

Does that mean now I will change who I am? Absolutely not!

Learning new things should always be adding to your existing trait. It should help build you into a better human, not a bitter human, with every experience. How do you want the world to taste? Let that sink in, you will automatically know how you should behave.

You might be thinking right now, "Of course, Anu!" Believe me when I say these things have to be done consciously; I am serious. When in pain and when we are facing an issue, it is very easy for us go into a dark place, especially a heavily traumatized experience. Some situations really do drain you emotionally, spiritually and physically. When that happens, make every decision thinking that whatever you do, and however you do it, it is for the greater good. If you have a little bit of lived experience, that's good! You will do better than earlier, but if you can find a mentor to work with, wouldn't that make your journey to success and learning much attainable? Not everyone can be a mentor. It's like networking, you need to find a mentor that best suits the goal you want to reach.

Are you thinking, "But how is this related to leadership?" I will tell you. When you are a leader, there are going to be unpredictable fires you will be putting out everyday and all the time. If you lose yourself, how will you keep a hold on your team? If you get demotivated, how will you keep your team motivated? Therefore, knowing your goals, networking in right circles and being comfortable with uncomfortable situations will help you a lot.

Today, along with my passion and goals, I also know how I function. I know what I am not comfortable with. Does that mean there are no more issues? Noooo! It means I have learned to handle them better.

When you go through metamorphosis, you are the only one in the cocoon. You really do get time to work on yourself and that is the key to all the learnings I wanted you to gather from me today.

Before I say "sayonara," I have just a few more thoughts I more often think about when I hear and observe things around me. I would like to share them, too. I think this type of mindset is very important for growth.

More than often, we have discussions about patriarchal society. It's essential to recognize that society is shaped by our definitions and perceptions. While protests and advocacy have their place, they shouldn't always be the first response. Instead, consider personal actions and self-reflection. If you've made a mistake, don't perpetuate it by treating others the same way. Understand why it happened and work on improving yourself. Remember, true change comes from within, and seeking help requires acknowledging areas where you need growth.

Another dubious discussion revolves around the imbalances we encounter in our lives. Contrary to the notion that personal and professional spheres are separate, they are interconnected facets of our existence. Imbalances will inevitably arise in both areas, but what truly matters is whether you have a voice, receive respect, feel heard, and believe your opinions hold significance. Reflect on your personal growth and consider whether your actions contribute positively to yours as well as others well-being.

I understand that this is a lot to take in all at once. Work on one thing at a time. Be gentle to yourself and do not be scared to take that leap. Sometimes most beautiful journeys are led by accidental decisions.

As an immigrant, I want to share to all my fellow immigrants when you start working (particularly in North America), you should familiarize yourself with professional etiquette, especially in relation to diversity and inclusion, because you are part of it. Learn how to navigate your experience and spread a positive impact.

Now, I would like to know, what is diversity and inclusion? What is it to you today at this moment? There are so many aspects that we overlook. Diversity and inclusion mean accepting an individual for who he or she is. How feasible is that in a professional environment? Let that sink in and give it a thought before you read my answer. For me in a professional environment whoever brings me a new idea or different perspective is contributing to diversity and anyone who is giving space to someone to grow is being inclusive. I believe it is that simple. Do you resonate with me? If not, that is okay. You know why? I would sit and listen to you and that is what welcoming diversity and fostering inclusion is. I hope I have made my point.

My dear friend, before I let you go, I want you to know that I have made my share of mistakes and it is not possible to have a professional or personal life without making mistakes. The key is to learn from them. When someone shares their struggles and how they overcame them, their strength and determination to stand up inspires me. From reading this chapter, I want you to feel lifted. Also, I want to mention one last time that not speaking up for yourself is the biggest mistake you will ever make. I am an advocate of speaking your mind to another person and I know that sometimes it is easier said than done. The key is to try. I had to work so much on myself to realize that listening to myself is very important. Listen to your gut instinct because everything else is just noises. Own your journey!

Before embarking on another incredible journey shared by my co-authors, I'd like to share one last debatable topic that I've often heard from many people that HR rules grant too much power, allowing exploitation, especially when it comes to women's safety rules. While it's unfortunate that such individuals exist, the purpose of professional conduct isn't to create discomfort—it is to prevent boundary violations. If you have clear intentions, driving these rules should not be an issue. They ensure workplace safety for countless employees. So, next time you question them, consider the situations that led to their implementation. This is another great trait of a leader. It is very easy to complain; instead, learn to understand.

When it comes to leadership in STEM, there is so much beyond academics. I would urge you to brush up on your emotional IQ, soft skills and network. Please put yourself out there and communicate, work with diverse sectors, engage yourself in volunteering. And if you want to be a leader, lead by example!

You need to find your tribe, whatever fits best, where you can find a safe place to express yourself, and you will find yourself moving ahead. If you cannot find that community for yourself, then make one for others. That's when you are a leader.

I hope this chapter has given you some fresh perspective!

Being a leader is challenging especially if you are a woman and an immigrant. From cultural changes, language barriers, and simply being heard, remember to be resilient and work towards your goal through determination. Own those roadblocks and pave your map to success; what is success to me might not be what it is to you. This is very important. Compare yourself with the previous version of yourself. Start there!

Anu Nair

Anu Nair, a dynamic individual blending creativity and analytical acumen, navigates the intersection of science and technology with fervor. Her academic journey began with a Bachelor's degree in Genetics, followed by a Master's in Biochemistry. Driven by a desire to create meaningful impact, Anu embarked on a multifaceted career.

Starting as a Biology Faculty, she nurtured young minds and instilled a love for learning. Her thirst for knowledge led her to pursue a diploma in Chemical Engineering from Cambrian College in Canada. In the world of chemical engineering, Anu honed her expertise in process control and quality assurance. She played pivotal roles as a Quality Inspector, Technical Lead, Research Formulation Chemist, Research Technician, and Research Facilitator contributing to diverse projects in construction manufacturing, renewable energy, and biotechnology.

Anu's commitment to excellence shines through her meticulous laboratory work and leadership qualities. Beyond her professional achievements, she holds a PMP certification, demonstrating proficiency in project management. Anu actively engages in community empowerment, by volunteering at various organizations.

Anu's journey reflects relentless pursuit of excellence, unwavering commitment to positive impact, and boundless curiosity. With a creative spirit and scientific mind, she continues to push the boundaries of innovation, contributing to a brighter, more sustainable future for all.

Connect with Anu at https://www.linkedin.com/in/nanu01.

CHAPTER 6

Embrace Yourself in Your Jungle

Brittany Overko

Do you enjoy seeing how things come together? Are you like me and intrigued by moving around the puzzle pieces to solve complex problems? After a three-year delay from COVID-19, I recently re-planned a trip to Alaska.

It took weeks of piecing together every activity in its allotted timeslot, but the result was an hourly vacation itinerary without a moment wasted. I was so proud of this masterpiece. Given this is how I enjoy my personal time, it may come as no surprise to you that my interests in logic, organization, and solving hard problems formed into a Science, Technology, Engineering, and Math (STEM) career.

My first exposure to engineering was in high school. I attended Miss Porter's, an all-girls school in Farmington, Connecticut. It was initially my parents' idea to go to school for four years without boys (not exactly my high school dream). But looking back, it was a life-changing opportunity as Miss Porter's shaped me in so many ways.

Instantly, my network exploded and I met the most amazing group of women. I worried less about what I was wearing or how cool my answers sounded. Sure, teenage girls can be tough to get along with, but here it was different. We were part of a culture that allowed us to grow our own voices, foster our own interests, and gain confidence in our individuality. There was a sense that we all belonged. We were the leaders of our own lives.

Miss Porter's is also where I experienced my first "Intro to Engineering" class that sparked my desire to become an engineer. The class was hands-on. We built bridges to explore how triangles are the strongest shape and programmed robots to follow lines on the floor during our coding lesson. The ability to bring math and science alive in practical applications and solve uniquely challenging problems excited me. While I had always excelled in school, my enjoyment had come from the satisfaction of an A on an exam. This engineering class was the first time I wasn't worried about memorizing dates or formulas for a test at the end of a semester. This class was different. It launched my love of learning and ignited a passion in engineering.

Fortunately, my learning environment in high school was very safe and supportive. You could try new things and it was okay if they weren't right or perfect. I thrived through the semester and finished that class knowing what I wanted to do in life. It was uncharted territory as there wasn't anyone in my family who was an engineer. My mom was in healthcare, my dad in finance and my brother joined the US Army after school. But as with any other new puzzle, I just had to figure out the pieces and make it work.

Things took a bit of a turn when I went to college. I pursued engineering at the University of Connecticut. Outside of the fun of competing on the equestrian team and following the basketball teams, classes were very hard. The first few weeks of my freshman lectures had hundreds of students in class, the majority of them male. As time went on and people began dropping out, it became more evident how few female engineers surrounded me. I quickly missed that network I had come to love from Miss Porter's. But I tried to keep my passion front and center, learning about different types of engineering and figuring out exactly what I wanted to do.

My first internship at Macchi Engineers LLC gave me a quick taste of how different the "real world" of engineering would be. At a job site meeting, I had the opportunity to shadow a Hispanic coworker. We sat down inside a mobile trailer at a large table. He leaned over as everyone was settling in and said, "Look around. You and I are the only minorities in this room." I stared back at him trying to register why he was considering me (a Caucasian) a minority. He then continued with, "You are the only female here, get used to it." This was becoming a common theme for me. I was the only woman in that meeting. I was the only woman that whole day at the job site. I quickly became very self-conscious. I started to wonder whether I had made the right career choice. Why did I want to be an engineer? Could I fit in and truly succeed?

Ultimately, I continued through my studies at UConn, soaking up as much knowledge and experience as possible. I continued through classes that were with mostly male students. I had additional internships with mostly male coworkers as well. At the end of my four years of college, I graduated with a dual degree in Management and Engineering for Manufacturing. I knew I still had a love for engineering that I wanted to apply to the world, so I started searching for the right role to launch my career.

I accepted a position with BAE Systems, an aerospace and defense technology company, as a participant in their Operations Leadership Development Program. This three-year rotational program allowed me to further explore career paths — taking on different roles annually at a variety of the company's national locations. During my final year in the program in 2015, I worked as an Operations Program Manager in New Hampshire. I oversaw product manufacturing and delivery while ensuring we met our cost, schedule, and quality requirements. I knew immediately that this was the job for me. The position was one giant, ever-moving puzzle where I had to constantly put the pieces together. I made sure the right resources were working on the right assignments and removed any hurdles that got in the team's way. I enjoyed the demanding environment. I could do technical engineering, but still be a team leader.

Everything was going great in this assignment, except for one of my Program Managers. He and I were more than different. He was challenging, never afraid to pound his fists on the table. When he wanted something, he was the loudest one complaining. I assure you, I am no shrinking wallflower, but I didn't quite have the same confidence I felt back in high school in the corporate world just yet. I was still learning to navigate a historically male-dominated industry and I wasn't always sure of myself. My Program Manager and I butted heads. Our team was behind schedule, and he wanted me to adopt his style – down the team's throats every day, forcing them to go faster. But this wasn't how I approached situations, and it made me even more unsure of myself.

A few months into this rotation, my direct manager asked me how things were going. I explained that I was struggling and didn't know how to be the leader I wanted to be while working with this Program Manager. My manager shared an analogy with me that changed everything and has stuck with me to this day. He said, "You're a zebra. You have stripes. Those stripes will never change. You will never have spots like a cheetah, so stop trying to pretend you're a cheetah." I thought about this for a few days. I would never have spots like the Program Manager wanted me to, but that didn't mean I couldn't succeed. Zebras and cheetahs both survive in the wild, but they do it their own way.

A zebra would never thrive trying to chase down prey and a cheetah would starve trying to live off the grass.

This pep talk gave me the confidence to stop trying to be someone else and to stop apologizing for it. I learned to be more direct and hold people accountable, but in a manner that I was comfortable with. I was able to lead the team in my own way. My confidence grew over the next few months. Yes, this led to more disagreements with said Program Manager, but I was making progress with my team. We finally did get back on schedule. I was unapologetically myself and I was hitting my stride.

I've worked as an Operations Program Manager for the last eight years, taking on various projects and challenges. Whether my coworkers around me are cheetahs, gorillas, or flamingos, we all have to survive together in this jungle. We all need each other to get the job done with our own sets of strengths. My jungle continuously demonstrates that there are people who want to help me and those that I can give back to as well. Yes, we may do things differently, but I've been able to learn from each person I interact with and grow from those experiences. My peers in high school bolstered my confidence. This new, ever-growing network in the office helped me regain that confidence and nurture my passion.

I shared before that my brother is in the military. At BAE Systems, our mission is – We Protect Those Who Protect Us ®. Producing aerospace and defense technology products that help our warfighters hits close to home for me. My brother and I got along growing up, but like most siblings had our differences. He favored history in school, while I went towards STEM. I never would have guessed we'd both eventually end up in the same industry. When my brother was deployed, I was scared knowing about the risks he would face, but I also knew that I could help bring him and others home safely. I could help our team ensure our products were the best they could be. Any job can be grueling and mine is no exception. Hard days can lead you to question, why am I doing this? But then I think of my brother and the role our team plays in protecting our men and women in uniform. That motivates me to take a breath, get my head back in the game, and keep doing what I love.

I'm 12 years into my career and I still have days that I feel like an imposter. I recently took the scary leap into a new position to continue my career growth. Now as a Program Manager at BAE Systems, I have an even larger puzzle to figure out as I lead engineering, finance, contracts, and other functions – not just the operations team. The sense of doubt and uncertainty has started to creep back as I learn the ropes with this new role. But I keep reminding myself to take a deep breath and trust myself. I know that I'm not the smartest engineer or the savviest business leader. But I still love what I do, the people I work with, and that motives me to tackle each day in my new jungle and keep learning.

Being an engineer has opened more doors than I could have imagined. I get a rush out of finding creative solutions, solving challenging problems and brainstorming a path forward with my coworkers. My journey into engineering wasn't always easy. But I followed my interests and found people who continued to support me along the way. I've learned that for those who stick with it, engineering opens possibilities and introduces you to many amazing people.

My fellow female engineers – you might not think you're smart enough to be an engineer or pursue a STEM career, but think again. You are good enough. You don't have to be your 'classic engineer' - the stereotypical person hunched over a computer all day. Or if that's what you do want, then give it a try. If you can find your own spark, what inspires you, then I have no doubt you can turn it into an amazing and fulfilling career.

So find what you're passionate about, cultivate a network that continuously builds you up, and never stop trying to solve those puzzles. Never forget to uncover your own spots, stripes, or feathers and embrace your amazing self for who you are.

Brittany Overko

Brittany Overko works at BAE Systems, a global aerospace and defense technology company. As a Program Manager, Brittany leads a cross-functional team through initiation, planning, and execution in accordance with specific program requirements, critical success factors and company processes. She works with her team to monitor ongoing work products for compliance to customer work statements, requirements, and quality assurance standards.

Brittany has worked at BAE Systems for 12 years. She graduated from the University of Connecticut with a Bachelor's in Management and Engineering for Manufacturing. After graduation, she moved to New Hampshire to start her career at BAE Systems in the Operations Leadership Development Program. Brittany obtained her Master's from the University of North Carolina at Chapel Hill. She holds an active PMP certification.

Brittany enjoys mentoring and coaching in/outside of work, helping to provide insight for others. She volunteers through her company with organizations like FIRST Robotics and Boys & Girls Club of Greater Nashua.

Brittany currently lives in Nashua, New Hampshire with her husband, Steven. In addition to work, Brittany spends time enjoying the outdoors by hiking or running, traveling with friends and spending time with family. She is also an avid horseback rider and scrapbooker. Brittany is thankful to everyone who has supported her in her personal and professional career.

Connect with Brittany on LinkedIn @brittanyjohnson15.

CHAPTER 7

❚❚ Press Pause

Claire Skillen

Throughout my 20 years of professional and volunteer experience, I've been perpetually learning, growing, and evolving. A few key building blocks have emerged as fundamental to my growth, whether serving in a leadership role or leading from other places. Here's what I have learned so far; I hope some of what I share resonates with you.

Know Yourself – Understand What Shapes You

Leadership can have many definitions depending on the individual, team, or type of industry. If you are newer to leadership in any capacity or have several years behind you, knowing yourself is a critical place to start—and revisit with regularity. So much of how we show up in the world and the choices we make stem from this. Ask yourself the question, "Who am I? How well do I know myself really? How connected do I feel to my authentic self?" Like me, you can be well into your professional journey and still commit to these reflections. Who we are is a process of becoming and unbecoming over time (inspired by Dr. Thema Bryant, author of *Homecoming: Overcome Fear and Trauma to Reclaim Your Whole, Authentic Self*), so we need more than a single snapshot of who we are to remain connected to an authentic self.

Regarding crucial areas to explore regularly, my internal core values hold a steady place on the agenda. Still, I also notice the things external to me, like activities, environments, people who give me energy, and those who take it away. And those are constantly changing. The better we know and understand ourselves at any moment and as we evolve, the clearer we can be on what guides us, how we function at our best, and how we recover when we miss the mark.

Working on myself and getting to know who I was first became a goal in my late teens. Having suffered from depression, I realized I needed to differentiate between who I am and what manifested as depression at the time. Without this distinction, any attempts at understanding my essence, my authentic self, would be muddied. It was an essential realization for navigating a world that continually pressures us to be other than who we are. I'm now completing a master's degree in counseling psychology, which has taken my self-knowledge and my knowledge of mental wellness to a new level.

In one of my graduate courses, I learned about the concept of a biopsychosocial framework (shown in Figure 1.), first developed by George Engel. This framework, which has since been updated to include spiritual factors, helped me understand how past and present biological factors (such as genetics), psychological factors (such as stress or trauma), and social factors (social infrastructure, social networks, upbringing, culture, and close relationships, to name a few) influence my values, my beliefs, and even my habits. This framework gives me insight into how I process the world around me and within me and helps me sort through the origins of my conscious and unconscious beliefs, views, behaviours, and biases. I spent a good while reflecting on this, and what I gained is unquantifiable: I developed a fuller connection with the fabric, colours, and design of my increasingly vivid tapestry of life and the areas I needed to work on and appreciate.

Figure 1. The Biopsychosocial and Spiritual Framework

Note: The examples in each category below are not exhaustive; there are many more examples and adaptations of this model.

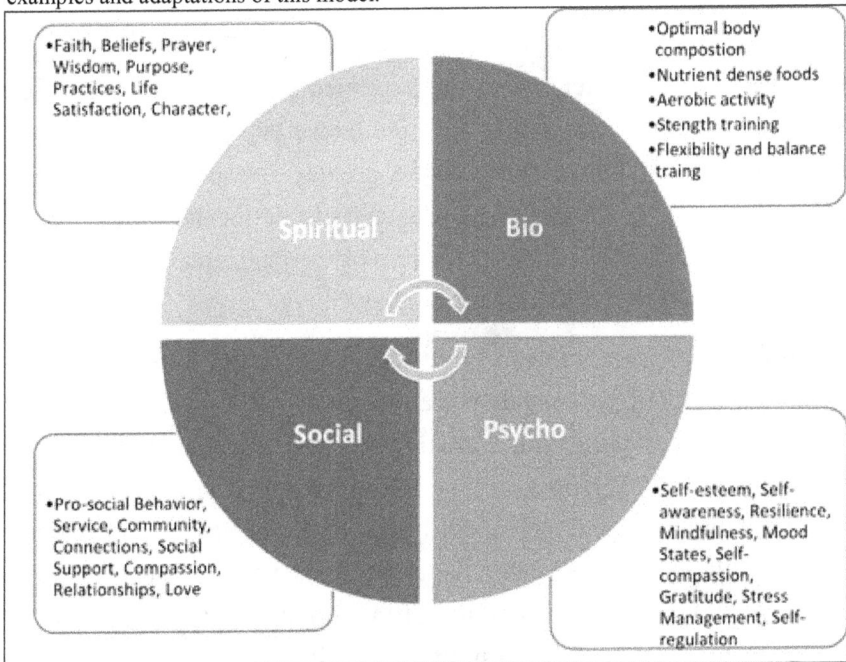

• Faith, Beliefs, Prayer, Wisdom, Purpose, Practices, Life Satisfaction, Character,

• Optimal body composition
• Nutrient dense foods
• Aerobic activity
• Stength training
• Flexibility and balance traing

Spiritual

Bio

Social

Psycho

• Pro-social Behavior, Service, Community, Connections, Social Support, Compassion, Relationships, Love

• Self-esteem, Self-awareness, Resilience, Mindfulness, Mood States, Self-compassion, Gratitude, Stress Management, Self-regulation

https://stephaniebryanphd.com/blogs/science-and-practice-of-health-and-well-

If you find yourself in a leadership role, knowing yourself on a deeper level helps anchor you in the turbulent times that could pull you away from who you are and what is truly important to you and your team. Building your self-knowledge can increase your openness to ideas, self-acceptance, and acceptance of and curiosity for others. In turn, that openness and inclusiveness will show up more consistently in how you show up and in the thoughts you choose to entertain.

How might biopsychosocial—and I would include spiritual—factors shape who you are, how you lead, and how you interpret the world around you and within you?

Know Your Core Values: Your Internal Board of Directors

I have learned to check in on my values, annually and ad hoc, to see where the alignments and misalignments are, depending on the situation. Sometimes, a significant life change can cause a value shift, and even little challenges can have a greater impact than you might realize. My five core values, Curiosity, Wisdom, Service, Authenticity, and Sanctuary, have remained relatively consistent; however, the weight or importance placed on each has changed.

I refer to my core values as my 'internal board of directors.' When something is out of sync, such as waning energy at the end of the week, body pain, or a feeling that something is missing, I try to map it to a misaligned value. Noticing this, identifying which value is out of alignment and why, and then identifying a path forward is a key part of leading yourself. In the past, I have experienced high levels of stress and burnout, which I attributed to putting too much pressure on myself or dismissing the signs by assuming 'this is the nature of the work.' My epiphany came on gradually: the physical symptoms and emotions bubbling up were related to a misalignment of values, a disconnect between my choices, and the priority I had placed on the values sitting on my board of directors. This important lesson changed how often I checked in on my values. I started to dive deeper into how my values influenced my choices and decisions, how I showed up in the world, and the impact this has on others and my life direction.

Do you know what your core values are? When did you last check in to see which values matter most to you? How are these reflected at home, at work, when leading yourself, or when leading others? What emotions, thoughts, or feelings emerged as you explored your values? How might your board of directors steer you to better align your values with how you budget your time and energy?

Flex Your No

I have always believed that time is our most precious resource. There are 86,400 seconds in a day. How many are you willing to budget for something? How do you allocate time with loved ones or friends, 'you time,' sleep, work, passions, hobbies, your favourite drink from your local coffee shop, or that extra 15 minutes in traffic?

I felt utterly stagnant twice: once in my 20s and another in my 30s. I craved the next milestone and wondered why I wasn't already there. While the reasons for this stagnation varied, a big part was not being able to account for all or most of my time and corresponding energy. As an A-type personality, I learned how important it was to audit my time, both how and where I spent it. When I sat down and committed to documenting my time activities for a couple of weeks, I got a more precise account of the gap between how I thought I was spending my time and the accurate picture. I evaluated this newfound insight against my core values and the overall vision for life to pluck out the inconsistencies and reflect on why I chose to spend my time that way.

One realization I came to was that I needed to flex my no. That's right. This tiny, two-letter word was my new power. I could set boundaries in places and spaces where I was uncomfortable and be confident that

I could manage the responses in all their forms. Some people find this easier to do at work than home, and others more at home than work. Some people—as was the case for me in my 20s and 30s—are working on boundaries across all aspects of life. The ability to set boundaries, recognize when they have been crossed, and communicate this to others is a foundational piece of leadership. When we set boundaries, we spend our seconds and energy on the things that matter most, as guided by our internal board of directors and organizational values, and we create a safer space for others to do the same.

I have noticed over the years that setting boundaries with yourself can be the toughest, as in many spheres, women are programmed to please others and place themselves last. Examples could be saying yes to assisting on a special project when you know you are already over the limit or staying up late to complete tasks when your body craves quality sleep. There are many more examples. Setting boundaries with others can begin with saying no to ourselves. The hard truth I had found in saying yes when no was the honest answer is that each time we don't honour our internal 'no,' we abandon ourselves in small and sometimes big ways.

How do you set boundaries at home, work, with friends, or with yourself? Are there areas in life where you can flex your no? What does it feel like when your boundaries are honoured? When they are not honoured?

Pressing Pause: Creating Space for Yourself in a 100mph World

Despite shifts towards wellness, society seems to continually prioritize doing over being, attaching significance and value to how much we accomplish rather than who we are to ourselves and others. A former coach once called me out, saying I spent too much time doing rather than being. I confess that I only partially understood what she meant at the time. I had been hurtling towards my second major burnout, and it hadn't occurred to me to audit my time or budget my energy. I didn't have it in me then to stomp on the brakes and allow for quiet introspection. My nervous system was ready for a mutiny, and the rest of my body sent warning signals for the impending collapse: new pains beyond the typical gym-induced muscle aches, increased mental fatigue, and intrusive thought patterns that required a prompt eviction from the mind. For many of us, it's impossible to press pause by retreating to a favourite tropical hot spot, nature sanctuary, or cityscape for six months as a reset. So, how do we stay connected to who we are and what we need in a world that constantly pulls us to be someone else or sacrifice wellness to complete that last task?

To escape burnout, I had to reframe what 'productive' meant, which included creating a better balance of being and doing. When I allowed myself to be a human 'being' and not just a human 'doing,' I regained access to those bright flashes of creativity, those real emotions that I had placed on the back burner because I viewed them as inconvenient at the time. I discovered my own authentic voice, clearer without the noise of the world around me, because I allowed myself to pause by allotting weekly blocks of time to let my mind wander and process. This wandering is where some of my best ideas come from, and it notably affects my mindset and resilience. I also needed enough space and time for my emotions to crescendo and retreat like waves on shore. Otherwise, I was suppressing my emotions and not regulating them. Some practices I now perform weekly without fail include sunrise journaling, post-work walks by the ocean, cold plunges, simple breathwork techniques for a quick 2-minute reset, and a 'priorities review' session that helps keep me focused.

What are your warning signs before your body's check engine light comes on? What are some things you do—or could start doing—to help you navigate the ups and downs and ebbs and flows of life? Are all the things on your task list a top priority? When was the last time you made yourself a priority?

Bringing it All Together

In my view, self-knowledge is at the core of what makes a great leader: know who you are so you know where you are leading from. How well we know ourselves influences vision and values, navigating change, advocacy, inclusion, innovation, and creating psychologically safe spaces. Understanding what shapes you, staying connected to your values, auditing your energy, flexing your no, and pressing pause are but a few of the building blocks for great leadership, and they, in my view, make for a solid foundation.

I hope you can press pause for a few moments, find your answers to the questions I shared throughout this chapter, and reflect on how these connect with your leadership style. What are your building blocks for leadership?

Claire Skillen

Claire Skillen is President of Ceascape Solutions; curiosity, service, innovation, and continuous learning and growth are some of the core values that have guided Claire's career over the past 20 years. She is self-employed in technology as a senior consultant through her company, Ceascape Solutions Inc.; she passionately supports women (including transgender and non-binary) in STEAM.

Claire is currently working towards her master's degree in counseling psychology, with a focus on complex trauma and client-centered therapies for depression, anxiety, grief, and loss. Outside of work and studies, living on beautiful Vancouver Island—the unceded lands of the Lekwungen People; you'll find her paddling her outrigger canoe, snorkeling, hiking, and taking photographs. Claire offers consulting services, coaching, and photography.

Connect with Claire at ceascapesolutions@gmail.com.

CHAPTER 8

Speak to Lead: Unlocking the Power of Communication

Diya Nair

Are leaders born? Or do we grow into leadership?

In my case, it has certainly been the latter. I'm not what most would call "a natural-born leader." Throughout my childhood, I was soft-spoken and shy, unable to even get a word out. However, since then, I have developed and grown step-by-step, learning to lead large groups and, most importantly, empowering others to become leaders by unlocking confidence and initiative in themselves.

When most people imagine the idea of STEM, they think of quantitative numbers, advanced technologies, and scientific theories. However, the foundation of STEM lies upon a crucial aspect that often goes overlooked: public speaking.

Throughout high school, I have been a passionate advocate for public speaking and youth communication. In the years prior to high school, even as early as 9th grade, I suffered from severe anxiety and fear towards anything involving public speaking—a sentiment most likely shared by many others.

In my final year of middle school, I founded a non-profit titled Learn, Perform, Inspire (LPI). Initially aiming to assist my younger sister and her friends, this organization was intended to help develop strong public speaking skills in our youth by utilizing the TEDx speaking curriculum in a fun and engaging manner.

Since then, over the past four years, I've been able to expand LPI, reaching students all across the globe, teaching over 500+ students whilst hosting over 13 full-length competitions in the community. In just a couple weeks of training, students as young as seven have been able to gain both confidence and conviction in the way they present and hold themselves. Every new session and every new season, I continue to see my young students transform from reserved to assertive, displaying the crucial skills that public speaking and communication builds. Many of the kids I teach dream of careers in computer science, engineering, medical sciences, and other STEM fields, and the skills they were taught through public speaking built the foundation for their dream and helped provide them the needed skills to confidently and articulately voice their opinions.

My experiences through teaching young students over the past four years has opened my eyes to the need for youth communication and public speaking and has built the foundation for the character I am today; it has allowed me to transform from someone who used to fear public speaking, to an advocate and coach for public speaking.

In the classroom, I naturally gravitated towards STEM. Math always came effortlessly to me and subjects like chemistry were always exciting, engaging, and interesting to learn about. Outside the classroom, though, I've dedicated a lot of my time to other activities such as public speaking, leadership, team building, educating and mentoring others. These are less common hobbies among STEM students, but I have found them highly fulfilling. The journey to becoming a leader in STEM is one filled with challenges, and even as a high school student who has yet to fully experience and navigate this growing field, recognizing the critical foundation that public speaking plays in this field is pivotal.

Specifically through the lens of female leadership in STEM fields, public speaking is intrinsically intertwined in effective communication, conveying complex ideas, inspiring and managing teams. Communication is the cornerstone for developing confidence and proficiency needed to lead diverse teams and build crucial skills integral in the STEM field.

Where communication has the capacity to enhance overall leadership effectiveness, its prevalence in modern society is only diminishing. The rise of digital technologies is only substituting face-to-face interactions, especially in our youth population. Social media platforms, while connecting people virtually, only lead to superficial interactions and a decline in meaningful, deep conversations. Additionally, the fast-paced nature of modern life and the increasing reliance on quick, fragmented forms of communication like tweets and texts contribute to reduced attention spans and a lesser emphasis on thoughtful dialogue. This shift is only furthering the current trend facing society today: lack of effective communication, which in turn leads to lack of effective leadership. 1

"The art of communication is the language of leadership."
~James Humes

Communication is a critical component of the modern workplace. Whether written or spoken, our interactions underscore each initiative. Yet, even still, many leaders struggle to define what constitutes good communication or the importance of it. Communication in its simplest definition is the ability to effectively share your thoughts, emotions, and understand others. Yet, innately, it is much more than that; your ability to communicate plays roles much deeper and intertwined in our daily life than just the simple action of talking.

I was able to see firsthand the remarkable effects of communication education on our youth, specifically its effects in boosting youth confidence levels and diminishing presenting anxiety. During online sessions offered through my non-profit, LPI, during the peak of covid, students were stuck in doors with little to no contact with other peers or adults. Yet, just a minute bit of public speaking coaching was able to completely transform my young students.

Eden, a past student of mine, rarely ever participated during any group Zoom discussions of team practices. During our 8-week course, we noticed her slowly gaining confidence, first engaging with small peer discussions to completely owning the room as she refuted point after point during our debate competitions.

Another student, Shrinidhi, went from never experiencing public speaking, to a star debater, now even helping us teach classes and conduct debate discussions with the younger students.

As you can see, it's evident that teaching kids communication and public speaking skills is transformational. I have seen time after time examples of its positive effects and now, more than ever, in an era riddled with technology and a society less reliant on themselves and more on their tech, do we need to open our eyes to the need for communication in developing crucial leadership skills.

Specifically in our youth population, communication is a skill that is only diminishing as time passes. As you might imagine, since 2012 the average amount of time teens spend with their friends has decreased by over 50 percent. In fact, the typical high schooler spends nearly nine hours on their phone every day—tweeting, messaging, and consuming endless digital content. Simply talking with and to one another is slowly being forgotten in our tech-driven era.

Today, our generation is more comfortable talking to machines than with each other. Finally, some states are recognizing the tragedies created by phone usage among youngsters and there is a bipartisan effort to crack down on rampant student phone use. States around the United States are passing laws, issuing orders or adopting rules to curb phone use among students during school hours. Will this completely solve the lack of communication problem? Unlikely. But it's a great start in the right direction.

"A high school sophomore confides to me that he wishes he could talk to an artificial intelligence program instead of his dad about dating," ~MIT Professor Sherry Turkle in the New York Times

---

It's very unfortunate, but the lack of communication in our youth is only further leading to teens struggling with social anxiety, which is resulting in the depressive downward spiral teens face today.

Developing communication skills in youth is crucial for expanding the next generation of leaders and preventing the negative consequences created by lack of communication. In an increasingly interconnected world, the ability to articulate ideas clearly, listen actively, and engage thoughtfully is pivotal for effective leadership. Youth who are equipped with strong communication skills are better prepared to inspire and motivate others, not only enhancing their ability to lead but also fostering critical thinking skills.

Moreover, early development communication skills help set the stage for future success in both personal and professional aspects of one's life. In educational settings, students who are able to communicate well have the ability to learn better by participating confidently in discussions, and advocating for their needs and ideas. These experiences help set the stage for a solid foundation for their future roles as leaders in various fields.

As they transition into the workforce, their communication ability not only helps distinguish them for future interviews or networking, but also empowers them to take on leadership roles with confidence. Understanding the need for development in youth communication is essential in allowing for the creation of capable and effective future leaders who can drive positive change in their communities and beyond.

Sadly, the way things are going, communication (rather than concrete, tangible skills) will be what separates the chaff from the wheat and will be what sets apart our future leaders. Communication has always been an important part of development, but the more advanced our technology becomes, the less skilled our communication becomes.

For current leaders in the workforce, effective communication is essential for building furthering leadership skills and serving as the foundation for both successful organizations, and leadership within them. The ability to communicate clear, articulate visions, goals, and expectations, ensures foundational organizing and efficiency skills aligned, helping prevent costly misunderstandings that may arise. When leaders communicate transparently and consistently, they foster an environment of trust. Employees are more likely to feel valued and understood, which boosts morale, enhances engagement, and fosters a culture of mutual respect.

Moreover, effective communication is crucial for change management. In today's rapidly evolving business landscape, organizations must constantly adapt to new challenges and opportunities. Especially when taking into consideration the current competitiveness in the job market—specifically in STEM—the ability to communicate your own individual skills is crucial in securing job opportunities. In an increasingly complex and interconnected world, the ability to communicate effectively is not just a desirable skill for leaders—it is a necessity.

Public speaking not only holds great importance in developing leadership skills early on in a child's educational life, but also in their future career and professional life. With over 75% of jobs requiring some aspect of public speaking, the ability to communicate publicly is a crucial skill needed in the workforce. Communication has the ability to create cohesive workforce environments, innovation, and increased exchange of ideas in jobs.

In fact, a study by the McKinsey Global Institute found that productivity improves by 20-25% in organizations where employees are connected. Effective communication tools and practices play a significant role in these improvements. Public speaking teaches the importance of conveying a message, listening and responding to an audience and receptive thinking, all of which are critical skills in the professional world.

Even with the ever imminent rise of AI in jobs, public speaking and communication with others still remains an ability untouched by technology, an ability AI will never be able to replicate. Where AI can perform tasks or solve problems that require human intelligence, it still fails to replace the human ability to create relationships with clients that serve the alignment of business and communications strategy. Public speaking still remains an essential skill in the workforce, a skill of utmost necessity in the age of AI, and a skill that every business and job will look for in their workers and their leaders. Contrary to popular belief and scare tactics, communication is a skill that can never be replaced by AI.

Dr. Mae Jemison, the first African American woman to travel in space, is a perfect example of this. Her ability to use her public speaking skills has allowed her to advocate for STEM education and increased diversity. Her engaging talks and lectures communicate the importance of inclusion in science and technology fields, helping inspire our youth—specifically those from diverse backgrounds—to pursue careers in the STEM fields. Dr. Jemison's ability to connect with her audience and convey her message passionately has made her a powerful voice in the scientific community.

With the example of Sheryl Sandberg, COO of Facebook (now Meta), her candid and persuasive public speaking has been pivotal in shaping conversations around gender equality in the workplace. Through her book, *"Lean In,"* and subsequent talks and interviews, she has sparked global discussions on women's leadership and empowerment. Sandberg's ability to communicate complex issues with clarity and conviction has made her a prominent advocate for women in leadership roles.

And for myself, last month I led and organized a full fledged official TEDx event in my community—TEDxEmpire Ranch Youth. Some of the kids, whom I first met when they were shy elementary students, took to the big stage. Having to direct sponsors, parents, volunteer coordinators, and most importantly, our youth speakers, was an enduring and laborious task.

What made it possible in the end was simply the ability to effectively communicate. The event went fantastically, and over a hundred parents, teachers, educators, and students joined us from around the city to celebrate and listen to the ideas of the young leaders of our generation.

My experiences through teaching young students over the past four years has opened my eyes to the need for increased communication and public speaking and has built the foundation for the character I am today; it has also allowed me to transform from someone who used to fear public speaking, to an advocate, coach, and mentor for public speaking. I will continue to strive to make a positive impact through effective communication and leadership—essential skills needed in society today.

As we continue into an era dominated by technology, the innate ability to communicate, persuade, and connect through public speaking remains an ability distinctly human. Instilling the need for effective communication in leadership affirms the irreplaceable role of human communication in a world driven by technological innovation.

Diya Nair

Driven by a passion for public speaking and community impact, Diya Nair (a high school student) is the co-founder of Learn, Perform, Inspire, a 501c3 non-profit focused on developing literacy programs and nurturing youth students in public speaking communication. Her journey has involved educating hundreds of students, organizing impactful local competitions, and raising significant funds for charitable causes.

In her current role as Global Giveback Campaign Manager, she leads efforts to provide school supplies and online education to girls in India, addressing educational disparities with commitment and innovation.

As the current Lead Organizer for TEDxEmpire Ranch Youth, she has successfully united local sponsors and youth speakers to create a transformative experience filled with innovation and change through communication. She is enthusiastic about bringing her experience in education, advocacy and global outreach to new challenges and opportunities.

Connect with Diya at www.learnperforminspire.org or through LinkedIn at www.linkedin.com/in/30diyanair

CHAPTER 9

Girls from Alexander, New York, Do Not Become CEOs

Emily Jane Carlson

I dedicate this book to my family. With their support,
all things are possible.

"Girls from Alexander, New York, do not become CEOs," is a direct quote from my high school guidance counselor and typing teacher. That was the supporting message that I received as a young adult about my abilities to be a CEO...a leader within any organization that I wanted to grow within. Those words are deafening to any girl, but especially to one who has high expectations of what she wants to become. Did I mention those words were spoken to me from a female? Truth is, I buried those words and really hadn't thought about them much until the past few years. However, as I've been focused on climbing the corporate ladder to obtain my high school goals, those words have become the fuel to skyrocket my career.

The definition of the word evolution according to dictionary.com is "the gradual development of something, especially from a simple to a more complex form" and I believe that the ways women have had to evolve in STEM leadership adheres to the definition in its truest form.

How has the role of female leadership evolved since I joined the workforce in 1991? It starts with the expanded opportunities now available to all women. In 2023, this topic has become a crucial performance indicator for many organizations. There is a significant investment in training to ensure everyone understands the importance of equity and inclusion within an organization.

In 1991, those opportunities were not readily available. As a matter of fact, it was quite taboo to ask the question of why a woman wasn't given the opportunity to be at that C-suite level. It was never asked why a woman wasn't up for the next big promotion and I certainly didn't know how to focus on it for myself.

After that high school experience, I did not immediately go to college. Instead, I joined the workforce and tried to determine the best place for me and what my next steps might be. At that time, I went to work for a small, local computer store where I learned everything from how to assemble computers, basic sales skills, order and vendor management … heck, even how to obtain a credit card machine for the store. With zero skills, I was given the opportunity to be immersed in entrepreneurship in a small town. To this day, I do not think I was meant to go to college immediately. After all, there's a lot to be said for trades and learning life's lessons before obtaining a structured education.

How does that lead to evolution in leadership? Because those skills that I learned at that little computer store are still with me today. And, with the ever-evolving framework of how people climb to C-suite, you need the exposure of being able to leverage real-world experiences as you manage teams and organizations. We have seen how giants such as Steve Jobs and Bill Gates have been able to build Global, multi-billion-dollar organizations without graduating from structured university education.

I was the only woman in our little town's technical sector to be a hands-on technical team member (not just administrative or sales) and I could talk and do those technical pieces that needed to get done.

In 1992 I became pregnant with my son (born February 1993 ☺). I didn't know exactly what that was going to mean, but that little guy came to the computer store with me every day in his pack and play. It was a great way to learn how to be a mom, but also how to manage an incredible amount of pressure. It may have been frowned upon at the time, but I decided not to stay at home and kept working until that little computer store shut down in 1997.

In 1998 I decided to obtain higher education (another step of my evolution as a leader in STEM) and enrolled my son in nursery school and myself in college. It was a big step for me. I wasn't sure I needed it but wanted to go after a career as a developer. Being one of three women in my technical classes, I didn't really have a lot of people to connect with. Add in a small child at home, and well, it was a much different experience than most of my high school class felt.

After college, the opportunities that were given to me were those that I was lucky to find. Not notably was the time that I was hired as a help desk support agent for a startup Internet Service Provider and had to work the night shift and some of those middle of the night calls (as you might imagine) took a different tune when they heard a woman's voice on the other end of the phone. I wasn't a 900 operator; I was a technical help desk support agent.

Fortunately, the umbrella of STEM leadership has changed drastically since the 1990s.

First and foremost, the world that women live in is much different, even more so Post-COVID. It is no longer a detriment that women have families that need supporting and that can mean that the 'where' of work being performed is now accepted to not be in the boardroom, but maybe at the kitchen table.

We must continue the conversation that you can be immensely effective if you are not in the office. Our world has changed, and this has changed how we support our teams. As leaders, however, we must be sure we are morphing our delivery so that we can command a virtual room. Confidence is hard for leaders, at times, but especially for women. Do your homework. Do your dry runs of presentations. Your evolution as a leader will come with the ability to deliver with confidence and commanding the respect you need and deserve, even from the dining room table surrounded by your kids' colouring books and puppies snuggling your feet.

Secondly, we must keep talking about how to grow one another via coaching and mentoring. We must move away from 'coaching' having any type of derogative connotation and understand the value that it brings to the people and companies where it is a respected practice.

We must also take time for reverse coaching and mentoring where a seasoned leader is paired with a younger team member thereby providing a different perspective. This is how we evolve the way we understand the younger generation and make sure we are truly supporting our new and upcoming leaders. While a seasoned leader in STEM brings their knowledge and skills from the past, a younger team member may bring new and fresh ideas about new technology. Both can benefit from each other. I promise this will bring you so much opportunity to evolve the way you approach work that you will want to take on additional responsibilities as their 'mentor.'

As a leader in STEM, we MUST invest the dollars needed to support the coaching and mentoring relationships. While some people see this as overhead and time spent away from the work others were hired for, it is the only way to keep management abreast of new ways of thinking and growing our organizations and ourselves. Think of mentorship hours as a new benefit that should be added to your packages.

Third, we need to TALK about the lack of women in STEM if we are going to evolve the marketplace and opportunities that are available. It can be uncomfortable, but we must look at hiring practices, pay rates and the way women are viewed in leadership roles in STEM.

As a young mom sitting at that computer bench, I had to seek my own opportunities and use shear elbow grease to move up the ladder. If I hadn't sought out the opportunities, they wouldn't have been provided to me. Nobody (and I mean NOBODY) was talking about evolving organizations and bringing in more women leaders in STEM. I paved my way.

Women should not be left needing to 'figure things out' on their own. We need organizations to talk openly about their hiring practices, how women can gauge traction in their careers, provide resources ... really dig into the Diversity Equity and Inclusion (DEI) practices that we know are developed.

I know it's difficult, but women need to accept help. Back in the early '90s we didn't ask or inquire about such things. But, as we have matured in many of our practices, we need to take hold and learn to ask for and accept help, which will enable you to rise to the level of position you are seeking. As John F. Kennedy said, "A rising tide lifts all boats." We must work together and be the leaders we claim to be.

Fourth, imposter syndrome is real. We need to acknowledge it to be true for ourselves. How can you raise the ranks in leadership if you are afraid that you aren't really as good as you are. Trust me, you are. You are THAT good and do not let anyone tell you otherwise. While this term is fairly new over the past few years, it has been prevalent in our society for as long as anyone can track. The way we enlighten our community of leaders is to show support and not be afraid to offer praise as needed. We are all too often criticize, which only continues to hinder one's feeling that they are not good enough to grow as a leader. What if, instead, we share with others our views of how good of a job they are doing? It changes the whole way we lead and grow in our roles.

Fifth, network, network, network. The doors and conversations that can start with a simple hello are paramount. Go to work in your community and find organizations that are looking for help to grow their reach. Through these organizations, you can not only grow yourself, but it is an opportunity to find those that have like-minded goals and achievements.

Look beyond your traditional networking opportunities; investigate your network and find something that interests you in a way that will allow you to grow as a leader. Board opportunities at charities and startups are great places to start.

Sixth, seek a reach opportunity. A reach opportunity is where you find a project that may currently be over your current skill set / position but allows you the opportunity to learn from it. When it comes to evolving the way we lead in STEM, we MUST give each other opportunities to go for assignments that are more aligned to where we WANT to go …and, not where we currently are.

Talk to your leadership and go for the reach! Explain the benefits of why you are the right fit for the assignment, how you will gain any perceived gap in knowledge or skills and the benefit of allowing you to be in a position or role that will give you additional growth opportunities. This is one of the most effective ways that we can really look at how we change and evolve the roles of STEM leadership.

When I was a high schooler with the glimmer of being a leader, I was brought down by someone who did not understand the importance of supporting one another (especially women supporting younger females). While I do believe that we've come a long way, we must keep talking in order to change and evolve who is leading, how we are leading, and the importance of growth when we do allow new viewpoints and approaches to leadership. Do not allow anyone to deter your potential!!

We must grow ourselves to a mindset that embraces the way we need to change so that we can set the groundwork for those we lead.

"Before you are a leader, success is all about growing yourself. If your actions inspire others to dream more, learn more, do more and become more, you are a leader."
~John Quincy Adams

Emily Jane Carlson

Emily Jane Carlson is a leading executive and mentor in the Healthcare IT consulting industry. Emily's 30-year career in the Information Technology and Project Management realms has focused on supporting her passion for the healthcare industry by delivering best-in-class quality projects to aid strategic initiatives and grow revenue for her clients.

Emily is a coach and mentor to organizations, the technology industry, and throughout her community. Her podcast, Powered by Authenticity, advocates to change the trajectory of equity and inclusion for females through insightful, inspiring conversations from women who are forging ahead in careers that do not have a balanced female representation. Emily travels the country as a conference and event keynote speaker focusing on topics ranging from her technical expertise to equity and inclusion.

Among her many accolades, Emily is most proud of being chosen in 2021 as one of the Top Women Leaders in Technology, Excellence and Innovation by Consulting Magazine. In 2022, Consulting Magazine awarded her Mentor of the Year acknowledging the unrelenting work she has done in support of other females.

Connect with Emily at www.poweredbyauthenticity.com.

CHAPTER 10

My Travelogue to Leadership

Felicity Coe

I am excited to share with you some of the details around my professional journey to leadership in the corporate world, and how the skills I acquired ultimately led me to build my own coaching business to help empower others. If I can share some tools and tips with you that make your journey easier or more impactful, then I'll consider this effort to be a win! With that in mind, in order to make things easy, in this chapter I am going to use this icon ★ to call out information that will hopefully be useful to you in your journey to grow as an individual and as a leader.

First stop: Getting on board

Prior to accepting a position at a defense contracting company, my husband and I agreed that I would be working to prioritize our 401(k) savings and focus on putting money away for retirement. The plan was to save well and work there for five years. I stayed for 20! What I didn't count on when I began my corporate journey was how many opportunities I would be given to learn, grow, and ultimately lead, and how passionate I would become about leadership and helping others grow to their full potential.

For the preceding decade I had chosen to do a different type of work so that I was able to be home more when our children were young. However, as the kids got older, I wanted to give my MBA a test run, and thus began my "career." I gladly accepted the opportunity for a corporate role, and during my tenure I had the opportunity to work in many departments in corporate including Finance, Operations, Quality Control, Communications, etc. To be honest, I didn't seek transfers to other departments, but on numerous occasions I was offered opportunities that stemmed from my curiosity around tools and projects, and my willingness to learn and investigate further. ⭐ *Curiosity will take your career to new places! Follow up on things that interest you. Take an interest in what others are doing. Ask questions. Learn as much as you can about a topic and become an expert!*

It was in the early days of internet when our company converted to SharePoint (a Microsoft product that allows organizations to create websites for storing, organizing, sharing, and accessing information) and I was hooked! I learned as much as I possibly could about the tool. Subsequently, when the company wanted to build an **Intra**net, my SharePoint skills and I were invited to join the Communications team to help create a companywide website so that everyone could collaborate more fully. I look back on those days and can't imagine how we existed without common repositories!

During my stint in Communications, I learned a lot about events and how to bring people together to plan and move forward. I was able to observe leaders in action, and the impact of face-to-face interaction and **experiential learning.** ⭐ *It is difficult to overstate the importance* *of hands-on learning, experience and practice. If people are able to create memories and feel what they are learning, they are more likely to be able to put that learning into practice.*

Next stop: Getting others on board

Although I spent many years in different departments, I ultimately found my "home" in IT, and that is where my leadership journey exploded. Although I have the typical computer and technical skills needed in business today, I am not a developer or an Engineer. I was immersed in IT tools and processes, so I was able to speak the language and provide support to the teams, but my proficiencies revolved around people. Using the skills I acquired while serving in Communications, I created a forum in which leaders and managers could come together to brainstorm, learn, reflect and strategize. We all know there are many components to a successful event, but off-site meetings can often provide participants with permission to fully participate in **collaboration and learning.** ⭐ *As leaders it is important to keep up with new trends. Making sure that you continue to grow in technical and soft skills is important, and it is also critically important to develop your employees by providing appropriate and interesting learning opportunities.*

Relationships are the backbone of our teams. When you bring people together to learn in an environment where they can have shared experiences, you open the door for deeper understanding and caring. Strengthening the trust between people working together is critical to success. ⭐ *Remember that relationships are built one conversation at a time. I encourage you to take the extra few minutes in the hall or on the phone to connect with a colleague on a level slightly deeper than the coffee or the weather. What you discuss is important, but how you make others **feel** is truly impactful. Do they walk away feeling belittled* *or energized? Heard or ignored? If you take the time to invest in others, it will pay off in spades for you and for them!*

Once I realized the success of any organization depends on its people, I began working to establish a bench of IT/ Business Operations leaders who would be able to **inspire a shared vision** and lead others to achieve identified goals. ⭐ *The ability to inspire others is a gift, and as a leader is a critically important practice. If you can get everyone on the same page and share a destination, it will be much easier to get where you want to go! To help inspire others, talk about your vision for the future, for your team and for the organization. Help others see the big picture. If you can confidently share your vision of a compelling future, you can inspire others to follow you there.*

The next phase of my career included creating and executing leadership development programs. We would begin by asking cohort participants a few questions:

1. Do you have to be born a leader? Are leaders born or made? While it is true that some people have a greater propensity for leading others, there are many leadership skills that can be taught. We have touched on some already and will unpack a few more. If you are willing to stretch yourself to learn and practice, over time you can acquire many of the skills you need to be an impactful leader.

2. What makes a great leader? ⭐ *Think of a manager or someone else you thought was a good leader. Did they inspire you? What behaviors did they exhibit that made you want to follow? Are these attributes something you can learn more about?*

Next stop: Revisiting

It is generally acknowledged that if you want to get better at something, you have to practice. Learning new skills and failing to put them into practice is a lost opportunity. This holds true for acquiring leadership skills as well. You have to practice active listening, practice having difficult conversations, practice prioritizing people, and the list goes on.

When you practice, you reinforce new pathways in your brain and that is why repetition is so critical. That said, it is not always easy! Hundreds of coaching hours and clients have shown me that changing our perspective and creating new patterns of thought and behavior can be difficult. For example, many leaders at all levels deal with Imposter Syndrome, the feeling that your talents and abilities are not good enough for the job at hand. Through coaching discussion, we evaluate and validate the experience and expertise that brought you to your deserved position. We practice replacing negative self-talk with positivity and use reality testing to boost confidence. It may be helpful to remember that in many cases women have been selling themselves short for a long time, so give yourself some grace as you practice these new behaviors.

Excursion: Women supporting women

Over time, as I continued to work on helping others elevate leadership skills, I realized that if you were trying to think of executive level role models it was difficult to find women to look up to. Feeling very strongly about this, I helped create the first company Employee Resource Group (ERG) to support women in their efforts to grow and move forward. Our first planning meeting began with a group of IT members, but word spread quickly and women from other parts of the organization wanted to participate. Women throughout the organization were eager to network and support each other personally and professionally. The idea was to provide support, education and encouragement to those wanting their career to move forward. Within a year we had Points of Contact in over 20 locations throughout the U.S. and hundreds of members. This spoke volumes: acknowledging and reiterating that women often must push harder to get their voices heard and require the support of others along the way. ⭐ *I strongly encourage you to **build a tribe of trusted individuals** that can help you on your journey. It is important for women to have others with whom they can share and bounce ideas around. Finding a few people to "have your back" can go a long way to feeling supported and respected, and potentially positively influencing company culture. For me, I found that diversity in those relationships (particularly age and gender) further enhanced the benefits for all of us; people with different perspectives can often challenge our thinking.*

A Souvenir

I already shared that leadership topics abound, but I want to leave you with something that might be easy to remember. Appropriately, the acronym here is L.E.A.D.

Listen: If you want to understand what others are thinking and what they have to contribute, you must really listen. This means that when you are in a meeting instead of thinking about what you are going to say, you need to focus on and process what others are saying. It means that when you ask someone "how are you" and they answer, you take the time to listen and process the answer. Sometimes people just want to be heard. It is good to remember that if you are overly concerned with winning an argument or having the right answer, you may not be fully present to hear what you are being told.

Enable others: As a leader, you want to support others in what they do. Find out (see "listen" above) what is top of mind for others and what you can do to support those you are working with. Are there things you can do (or not do) that will help them move forward? Be a Connector: introduce people who can support and help each other.

Advise: Although it is typically good practice to ask questions to clarify a situation, as a leader it is often your job to offer guidance when needed. If you are the expert in the room, don't underestimate or second guess your skills; be confident in sharing your expertise. For some people, this takes practice and repeated positive self-talk. Remind yourself of your accomplishments and experience.

Delegate: As a leader, you will have a lot on your plate. Allowing others to "do" will not only help you get more done, but it will provide others with the opportunity to learn and grow.

Final Leadership Destination: Professional Coaching is my jam

The pandemic brought about many changes, and it was during that time that I decided to move toward my passion for coaching. In addition to providing valuable time to think and process, a coach can help you gain new perspectives, see blind spots and test out ideas in a safe space. As a Certified Professional Coach, I have the privilege of working with people who are navigating their careers and leadership journey.

There are so many components of leadership, and it will be up to you to determine what skills you want to develop and what areas you want to focus on. ⭐ *Think about what behaviors you want to exhibit as a leader and make note of it, so you can refer back to your list over time. At the end of this chapter, I will share my Leadership Topics list which I continue to expand. However, I encourage you to create your own list so you can be sure to include what is paramount to you!*

As a coach, I utilize various assessments. They are not always necessary, but often an assessment can play a crucial role in coaching, as both the client and I gain valuable insights into their strengths, areas for growth, and progress. Based on results of an assessment, a tailored approach can be created to help address unique areas of development.

If you get a little stuck trying to decide on what areas of leadership to focus on, an EQ-i assessment might be helpful. EQ-i is a measure of **emotional intelligence** and can help you understand how you behave in various situations (stress, etc.) as well as how you handle your emotions. Some topics addressed in EQ-i are self-perception, self-expression, interpersonal, decision making and stress management. Increased self-awareness in these areas can be critically important as your behavior can often impact others on your team, customers, etc. ⭐ *I encourage you to consider taking an EQ-i assessment as it can be extremely useful in pointing out areas of strength as well as where you may have opportunities for improvement.*

There are dozens of assessments available for furthering self and/or team awareness and in some cases free resources are available. I have found the **DiSC assessment** to be useful with teams as it provides insight into the behavioral style and personalities of individuals on a team; it also provides a common language. If you are interested in assessing your leadership practices, I am a big fan of the **Leadership Practices Inventory (LPI), a 360 assessment** that helps you better understand how you see yourself as a leader as well as how others see your leadership skills.

Create your own Itinerary

One of my favorite parts of working in leadership development and coaching is watching people grow! When someone is "all in" and is willing to be curious, and then follow that up with practice, the results can be tremendous. Trying to see things from a different perspective, and experimenting with new ways of doing things can be scary, but that is how we grow. ⭐ *The challenge is learning to be **comfortable with being uncomfortable**. Feeling fear and discomfort, doing "it" anyway,* *and moving through it.* Below is an image I love to share which reminds us that growth happens outside our comfort zone.

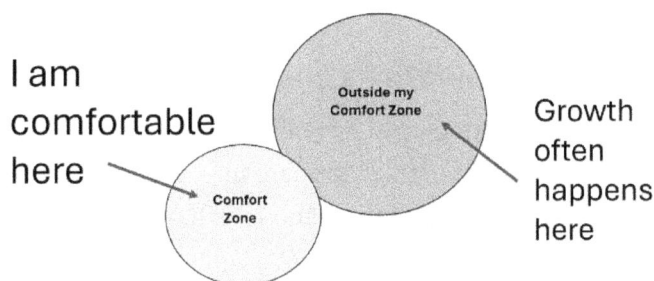

I am comfortable here

Comfort Zone

Outside my Comfort Zone

Growth often happens here

As a woman in leadership, I hope you find the tips below useful. Remember you may occasionally need to take an alternate route!

1. Don't forget to **ASK** for what you want. More often than not, women don't ask for the raise they deserve, the job they want, the projects they want to work on. Don't count on someone recognizing your good work and rewarding you for it. Promote yourself! Share your accomplishments and verbalize your goals. You may not get it all, but you certainly won't get it if you don't ask for it.

2. Don't be afraid to set **BOUNDARIES**. If you are given work that doesn't align, use your voice to question it. Sometimes it's ok to say no. There will always be situations where we have to do things we don't thoroughly enjoy but take the time to evaluate the big picture.

3. Get a **MENTOR** or a **COACH**. Many leaders are offered this opportunity but, if not, you can request this. If your company is not willing to pay for a coach, look for a leader who inspires you and see if you can collaborate. Seek out the expertise you are looking to develop.

4. Don't underestimate **SELF-CARE**. I know you have heard it all before, but taking care of yourself in order to better care for others is good advice. Set aside time for self-evaluation and make sure you are getting the support you need: physical, mental and emotional.

I wish you tremendous joy in your growth journey! If you would like a co-pilot or thought partner to share in your journey, check out my bio section for contact information and let's have a chat.

Safe Travels!

* My list of LEADERSHIP TOPICS (always growing)

- Accountability
- Adaptability
- Authenticity
- Challenging status-quo
- Collaboration
- Communication
- Conflict Resolution
- Decision Making
- Delegation
- Developing Employees
- Empathy
- Executive Presence
- Flexibility
- Inclusion
- Innovation
- Inspiring Others
- Interpersonal Relationships
- Leading through change
- Mental Health
- Burnout
- Negotiation
- Problem Solving
- Productive Feedback
- Self-Actualization
- Strategic Planning
- Stress Management
- Teamwork
- Time Management

Felicity Coe

Felicity Coe is the Founder and Principal of Coe Consulting, LLC. She is a Professional coach, certified by the International Coaching Federation (ICF) and has a master's in business (MBA). Prior to coaching, Felicity garnered two decades of experience working in various departments of a defense company, including Information Technology, Finance, and Communications. Working with the CIO and Business Operations leaders, she created leadership development programs that supported organizational development and succession planning.

Felicity has spent hundreds of hours coaching Sr. Directors, Directors, Sr. Managers, and Managers empowering them to move forward with personal and professional goals. She works with leaders on becoming more strategic to increase productivity and personal satisfaction. Felicity also helps individuals identify and overcome Imposter Syndrome, increase self-confidence, and develop skills and tools that will help them define and achieve goals. She is certified in a number of assessments including DiSC, EQ-i and Leadership Processes Inventory (LPI 360) which she utilizes with clients when appropriate.

Connect with Felicity at www.CoeConsulting.net.

CHAPTER 11
It Takes Courage to Fail

Heimy Lee Libu Molina

I dedicate my chapter to my parents, Wilfredo and Amalia Molina, my siblings, Wilson, William, and Anjanette Molina, with my sister-in-law Robin, my mentor, Elinor Moshe, and my fiancé, Tony Banning. With all of you, I have the courage to fail. With all of you, I am triumphant.

Failure takes courage, winning takes work—for an individual to attain success, one must fail. As contradictory as it seems, failure is a part of any victory. Without failure, one does not learn. Without failure, one does not know how much they want the goal. Without failure, no one enhances their courage. For one to attain victory, one must put in their best work and be ready to take on any level of challenge. No one gets to their optimum level without failing. The people who overcome challenges are the ones who initially failed in it.

It might be hard to overlook failure and become optimistic about it due to the stigma it holds, especially as women. In a personal sense, and as mothers, people can get immensely critical about our choices and actions. In a professional sense, women in management are more critically observed by their team. As leadership roles are traditionally dominated by males that adhere to convention, female leaders are often judged as either too soft or as an iron fist. With these assumptions that surround our society, it's easy to assume that failures are nothing but negative.

How hard is it to embrace failure? That depends on how ready you are for a breakthrough that requires "Failing positively." Impostor Syndrome is one of the most common, yet one of the hardest to defeat, "mindset killers" of our time. As a former "toxic perfectionist," I suffered Impostor Syndrome tremendously. One instance was during my year 5 at school. I badly wanted to become the best in my class, but the fear of not being enough and being mocked when I made mistakes was even greater than my urge to be the best.

I wanted to become the best without making any mistakes, so I did not take any risks—which means I did not take any significant actions, which means I did not do my best. On top of that, the towering feeling of not being suited for the highest rank of my class consumed me. Even with all that, I still hoped to be the top 1 of my class despite my shortcomings. To no surprise during the results release, I was not the first, nor the second, nor the third—I landed in seventh place.

Upon hearing this news, I was filled with regret: the regret of doubting myself, the regret of not pushing myself to take the risk and make mistakes, and the regret that I did not realise this soon enough. That experience taught me a significant lesson that I carry with me today—the feeling of regret is so much worse than the feeling of failure.

I wish that I had taken a lot of risk and didn't care about failing or what other people had to say. I wish I didn't think that I wasn't enough or that I wasn't born to be the best (for the best people are made, not born). I wish I didn't let the fear of failure and Impostor Syndrome fully control my decisions.

The acceptance of this fact was, and still is, an iterative process, so be kind to yourself and allow for the whole acceptance development to unfold into your life. This type of change does not happen overnight and within one instance. However, this can be fast tracked quite easily. To expedite failure acceptance, you must embody the following concepts in life:

We all have the perfect track record of surviving our worst.

This is slower   as compared to this.

One of life's greatest paradoxes is that the ups and downs of life take you further and faster than a monotonous and linear life.

Although consistency is key in most cases, there are some things in life that are beyond your control. The inevitable presence of failure is one of them. Once faced with one, we must assure ourselves that we have the perfect track record of surviving the hardest things in life—and that includes your mistakes, your failures, your blunders.

The version of you that you are currently experiencing is the product of all your wins and failures, which means that you are a living testimony of your resilience. Be kind to yourself, for the failures that you thought would break you, are the failures that made you.

Similar to the image shown above, the dips of life expedite our journey towards our destination, our goals. This is mainly due to the fact that there is momentum built up when we descend, and this is true in both physics and in life—when you experience a situation that brings you down, remember that this experience is preparing you for the incline that you're about to embark on. Remember that this experience allows you to gain momentum, to become an even better version of yourself once you reach the higher side—making you ready for the deeper dips and the higher inclines. Remember that this experience is a product of our resilience, our courage, and the decisions we made despite failure.

Always remember that life is 20% what happens to you and 80% how you react to it. You are not defined by the number of times you fall, but rather, the number of times you got back up. Every single individual in this world is bound to have a failure. The fine line that differentiates the people who are stuck from the people who are moving forward are our choices. So make your choices count.

Failure is the best teacher you'll ever have.

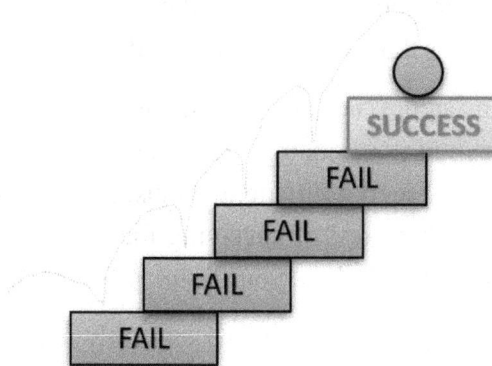

Failure is a two-way street: If you treat it like a negative thing, it will start treating you negatively, and vice versa. How can failure choose which side to go on? It can't, for the choice is only held by you. A moment of failure can either jumpstart your motivation to get to your chosen destination, or it can be a trigger for a downward spiral. The power to choose lies deep within you. Always remember that failure is not the opposite of success, it is part of it.

There's a lot more to learn when you lose rather than when you win—and that is why failures have the potential to drive you to success, as contradictory as it may seem. Take my previous story, for example. Because I did nothing but sulk and got into the "victim mindset," I did not achieve the goal I wanted. If I did the exact opposite and used my mistakes for the better, I would have shaped myself into becoming a better person, one that is worthy of first place.

This regret of taking the wrong path, will I let it define my next steps? Will I let myself go in a downward spiral because of one missed goal? Will I let this failure define me? Of course not! That failure turned into drive, resilience, and hunger for victories. I chose to use that failure for good, to use it as a steppingstone toward my next success. Now, I want to ask you, are you going to let a mistake define your path in the coming years? Or will you use it as fuel to drive your dream into fruition? The choice is yours and yours alone.

Make peace with fear; make friends with failure

One of the biggest misconceptions there is about life is that being courageous means you are fearless. Courage is not the absence of fear, but is the ability to act in spite of it. The ability to process failure enhances your courageousness. Once you've wholeheartedly accepted this, you'll make peace with fear. You'll treat fear as if it's a "driving agent," something that keeps you motivated. It's not an overwhelming being that can make or break your life; it is a part of life that you have accepted, and it's there to induce courage throughout your day to day. I know it will be easier said than done. As with failures, the bigger fears are, the harder they are to swallow.

Not all failures can be easily counteracted, and some of them are out of our control. If we accept the fact that things that are out of our control, they cannot get under our control no matter what we do and accept failure's presence in our lives with courage, we'll definitely make friends with failure. Making friends with failure does not necessarily mean being a failure, it means you've accepted its presence and know how to deal with it.

Remember that "If it's out of your hands, it deserves freedom from your mind." As I have repeatedly said in this chapter, be kind to yourself and life will repay you with the same kindness hundredfold. One practical habit that I'd like people to get into is "Apologise less, thank more." This is a simple, yet practical and effective way to slowly have the courage to fail.

Apologising puts a FULL STOP

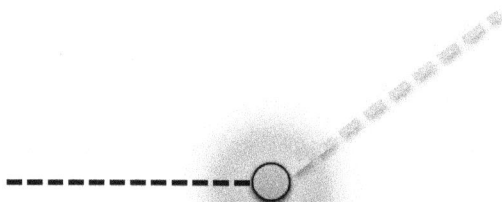

Showing gratitude encourages GROWTH

It's simply replacing the negative annotation that usually comes with mistakes/failures. An example is that if someone points out a blunder you've done at work, instead of saying "I'm sorry for making that mistake," say instead, "Thank you for flagging this" or "Thank you for pointing this out." Instead of saying "I'm sorry I'm late," say "Thank you for your patience." Once you've shown positivity in dealing with failure with your words, your actions will follow. Incorporating these tiny habits in your daily life will have a compounding effect in the years to come.

When struck with the fear of failure, accept that it is a part of life and give yourself assurance that what matters most is how you overcome it. You'll come out as a stronger and wiser version of yourself in the end. Once you've accepted the presence of failure and have learned how to deal with it better, you'll eliminate the fear and you'll be left with nothing but self-development. There would be more room for success in your life, which will leave you no choice but to throw doubt and disappointment away from your journey. You have no choice but to "fail positively" each time.

Being an immigrant, I felt like my journey from the Philippines to Australia was filled with failure after failure after failure before I got to the successes. I felt like I was on the brink of giving up most of the time. But my realisation in my previous story during year 5 hit me, what do I choose? The feeling of regret, or the feeling of failure? And this time, I chose the second one.

At the age of 19, I went to Australia with my sister, bringing nothing but hopes and dreams. These things fuelled not only my desire to excel, but also my urge to not succumb to failure, for I know that these would happen a lot during this significant time in my life. In my first few months, I had a lot of failures right off the bat: failure to recognise an Australian accent, failure to adapt to the local curriculum in Australian universities, failure to speak English fluently, failure to fathom whether or not I can survive my engineering degree, as I almost failed my first physics quiz.

It all felt so overwhelming as it was hard to find the positivity among the chaos that had befallen me until one day when I listened to one of Will Smith's interviews where he said that the best things in life are always hidden in the other side of terror. That thought has motivated me to see these failures as steppingstones toward my success. Little did I know, I started working on these failures little by little, one after the other. The desire to succeed, the desire to become the best version of myself, and the desire to be resilient for my family back in the Philippines overpowered my fear of failure. And now, here I am, writing this book, sharing my story to inspire you to have the courage to fail and use your failures to skyrocket your success journey. Life is good and I am truly grateful to be able to share this knowledge with you.

After I applied the "failing positively" mantra in my daily life, I became free from the shackles of convention, self-doubt, and fear of failure. It was all onwards and upwards from here on. Because of this, I have managed to attain various accolades that I did not even know I could attain. I have experienced a breakthrough that I did not foresee, and I have unearthed the best version of myself.

Now, I am not only a full-time engineer of a big project, but I am also a business owner of a personal branding and consultancy company called Build Your Brand (BYB), where I assist individuals and groups in skyrocketing, not only their present career, but also their career prospectives in the future. It is a business where I do not "fix" what's broken, but rather unveil the best version of people, which, in turn, unveils the best path of their future in a personal and professional sense.

I am also a committee member of Young Engineers Australia (YEA), a co-author (of now two books), a speaker, and most importantly, a mentor. Outside of my career, I venture out to various schools and universities in Australia to mentor kids and to show them that female presence in a male-dominated industry is not a wrong image to portray and that there is nothing more rewarding than being able to attain the goals that you have set yourself. The articles and podcasts that I have been featured on also contain the same message, which can be viewed through my LinkedIn profile.

The collective impact of all these actions has resulted in more prominent milestones that I have recently accomplished. Some of these are attaining the prestigious 2023 and 2024 7News Young Achiever Awards Finalist and People's Choice Awards, Women in STEM Education Engagement Champion 2022, Vice Chancellor's Excellence Award in Engagement and Sustainability 2022, Engineers Australia's Emerging Professional Engineer of the Year Nominee 2023 and 2024, and many more. None of these awards would have come to fruition if my fear of failure governed my decisions throughout the past few years.

Dear Reader, I believe you are capable of achieving the same level—if not higher or more—milestones in life. I challenge you to reject the urge to fear failure, for I know that deep inside you are filled with courage. Let this courage dominate you and the decisions you make in life. Expand your network, connect with like-minded people, read other people's stories and learn from them. Immerse yourself in this culture of courage and resilience by taking the first, tiny step—and the next step, the next step, and the next. Soon you'll find you have already built an empire for yourself. So, are you ready to unleash the best version of you? Only you can answer this question.

Heimy Lee Libu Molina

Heimy Lee Libu Molina, originally from the Philippines, is a Civil Engineering (Honours) graduate at Western Sydney University with First Class Honours and the University Medal. She is currently a Graduate Engineer at Gamuda Australia, working on the Sydney Metro Western Tunnelling Package and the Founder of Build Your Brand (BYB), a coaching business for personal branding and professional development.

Heimy is also a committee member of Young Engineers Australia, a co-author of Secrets of the Construction Industry, a Nominee of Engineers Australia Emerging Professional Engineer of the Year 2023 and 2024, Winner of the Vice Chancellor's Excellence in Engagement and Sustainability Award 2022, Emerging Designer of the Year Award 2022, Women in STEM Engagement Champion 2022, a Finalist for the 7News Young Achiever Awards in 2023 and 2024, and a People's Choice Winner for the same award in 2023 and 2024.

Aside from this, Heimy is actively engaged and invited in various schools, universities, panelists, podcasts, and events to speak about her resilience and to exhibit her contributions to the field of Female Empowerment in STEM and other male-dominated industries.

Connect with Heimy at

https://www.linkedin.com/in/heimy-molina-385a97217.

CHAPTER 12

Beyond the Glass Ceiling: Navigating STEM Leadership as a Woman of Color, Immigrant, and Mother: Embracing Challenges and Transforming Stereotypes into Strengths

Juanita Ama DeSouza-Huletey

This book is dedicated to my parents, Ernestho and Fidelia de Souza, whose love and support have been my foundation. It is also for my husband, Emmanuel, and our three sons, Ronald, Ernestho, and Pete, whose encouragement and belief in my dreams have been unwavering. Additionally, this is for all women in STEM with intersectional identities, whose courage and resilience light the way for future generations

Introduction - A Triple Challenge

"In every challenge, I found strength and the seed of opportunity. My identity as a woman of color, an immigrant, and a mother has enriched every step of my journey. Through these experiences, I discovered my purpose and turned challenges into steppingstones. My journey in STEM leadership reflects not just my career but the very core of who I am, leading me to places I had never seen before."

In my 32-year journey through the tech industry, I have learned to embrace challenges as opportunities, a mantra that has guided me through countless systematic and invisible barriers. As a mother to three wonderful sons, I have faced unique challenges in the world of Information Technology. Despite the slow progress of reforms, particularly in addressing intersectionality, I have taken it upon myself to share my experiences and serve as a catalyst for positive change within the industry.

Growing up in Ghana was marked by being a girl who didn't conform to traditional expectations. I was labeled a nerd, bullied, and subjected to derogatory names because I loved to "figure things out" and engaged in activities typically associated with boys. My father, the late Erenstho de Souza, was the only one who truly understood and supported me. I achieved double honours in Computer Science and Economics, becoming the first female graduate with that combination. My journey then led me to Winnipeg, MB, Canada, for a year-and-a-half internship with IBM Canada Ltd, where the rollercoaster of challenges began. Yet, I viewed each obstacle as a learning opportunity, to learn and grow, constantly defying my own expectations.

Now, I find myself immersed in the complex landscape of STEM, specifically in Information Technology and Data Analytics, where innovation is constant, and the need to 'figure things out' prevails. Being a woman of color gave me a unique perspective, turning challenges into steppingstones and biases into bridges, shaping me into the leader I am today.

Beyond the Glass Ceiling - Breaking Stereotypes in STEM

The 'glass ceiling' in STEM symbolizes the invisible barriers that prevent many talented women from reaching the highest echelons of leadership. I have personally felt these barriers, even as I ascended to the role of divisional head of an IT department serving 34 organizational units and over 1800 customers. My journey was further complicated by the triple challenge of being a woman of colour, an immigrant, and a mother.

I refused to conform to the expectations and stereotypes imposed upon me, as I've faced biases and discrimination, struggling to be recognized for my skills and competence rather than my skin color. There were even instances when I was mistaken as a clerical assistant rather than a leader, highlighting the deeply ingrained biases. Language barriers and accents also added to the complexities, with some dismissing my ideas simply due to my accent. Despite these challenges, I saw them as chances to prove myself and push beyond stereotypes.

The nuances of being an immigrant added complexity to my journey. Adapting to new cultural and workplace norms while maintaining my identity and values was tough. I felt like I was climbing a mountain without a map, always trying to catch up to those who had a head start, but my immigrant perspective brought fresh insights and innovative solutions.

Balancing motherhood with a demanding STEM career presented its own challenges, forcing me to make tough decisions about work and family. I often missed networking events and social gatherings that are key to career advancement. However, motherhood taught me valuable life lessons, which I've applied in my leadership roles and passed on to my adult sons now navigating their own paths.

Intersectionality - A Unique Perspective

Being a woman of color, an immigrant, and a mother has intensified my challenges but also endowed me with unique perspectives. My diverse background has proven invaluable in the STEM field, providing a range of viewpoints, insights, and innovative approaches to problem-solving. Continuously proving myself, I've had to work diligently to transcend stereotypes.

Motherhood has instilled in me patience, multitasking, and determination—skills that have been integral to my role in STEM leadership. Despite the resilience of the "glass ceiling," each effort I make not only tests its limits but also smooths the path for those who will follow. My role as a woman of color in STEM is not just about representation; it symbolizes hope, potential for transformative change, and the promise of a brighter future.

Ultimately, the obstacles I've faced and the achievements I've garnered extend beyond my individual experience—they resonate with every immigrant striving to overcome barriers and realize their dreams in STEM. My journey invites others to reflect on their own stories and realize how the power of their narratives can motivate broader change and deeper understanding. In a field fueled by innovation and problem-solving, the inclusion of diverse voices and experiences is essential for its progress.

Shattering the Glass Ceiling - Persistence and Resilience

Despite the persistent 'glass ceiling' in STEM, I remain dedicated and resilient in my pursuit of leadership and transformative change. I recognize the barriers but focus on breaking them, not just for myself but for future generations. My reflection in this metaphorical glass serves as a powerful reminder of my identity, my journey, and the inner strength that drives me. Each challenge I face reaffirms my commitment to shattering these barriers, ensuring a path that is more accessible for those who follow.

Breaking through the barrier demands daily effort and steadfast determination. Every push forward brings us closer to a more inclusive and diverse STEM community. I aim to challenge established norms, advocate for change, and serve as both a role model and mentor for future STEM leaders. My journey transcends personal success—it's about revolutionizing STEM for everyone. I envision a future where the "glass ceiling" is completely shattered, making opportunities accessible to all, irrespective of gender, ethnicity, or parental status. By sharing my experiences and insights, I hope to inspire others to join in this collective effort to redefine what success looks like in STEM.

Breaking Down Barriers

Throughout my journey, I refused to be limited by societal boundaries despite setbacks, such as promotions based on favoritism and unresolved

HR issues. A revealing conversation with a new executive made me more clearly see the deep-rooted biases and systemic issues in tech leadership. This wasn't just a personal setback but a wake-up call to fight harder for change and fairness.

I realized that the lack of advancement opportunities in tech wasn't a reason to give up but a reason to act. My journey was about making it easier for future generations of women and girls to reach top leadership roles, hoping to leave a legacy that helps them face fewer obstacles and have more chances to succeed.

My Leadership Voyage

My journey to STEM leadership has been like climbing a steep mountain, full of challenges and victories. It started with my role as an IT supervisor, where I had to overcome doubts about my "supervisory skills." Despite meeting all the criteria, I faced skepticism and had to really prove my abilities through successful project management and leadership initiatives.

As I ascended, the challenges grew, similar to facing harsher conditions on a mountain. I encountered numerous workplace obstacles, biases, and systemic barriers. Yet, I understood the importance of forming connections with fellow climbers—my colleagues. I made an effort to engage with them over coffee or lunch, as they became my support network, offering advice, lending a sympathetic ear and helping me navigate these challenges. This support network proved invaluable as I continued my ascent.

Reaching the top as the Divisional Head of IT and a member of the Executive Management team was both rewarding and isolating. I was often the only person with an intersectional identity in executive meetings, where every decision of mine was heavily scrutinized and the margin for error seemed smaller than ever. There was an unspoken expectation for me to outperform my peers to justify my seat at the table.

I also often found myself as the sole advocate for positive changes for initiatives that championed innovation, continuous improvement, better hiring practices, and transparency. The pressure was immense, for I was not only representing my department but also everyone aspiring to be in a similar position.

Addressing both overt and subtle biases added another layer of complexity. Microaggressions such as, "You're not like other women in your position," or, "It's surprising how well you manage such a technical role," were not uncommon. These comments reflected preconceived notions about my capabilities due to my gender and ethnicity.

Juggling my identity with the demands of my position required a delicate balance. On one side lay the weight of institutional legacy, and on the other, the hopes and dreams of countless women of colour who looked up to me as a beacon. Nevertheless, with each challenge, I remained committed to my mission—to ensure that *this climb wasn't just for my own success but to create a legacy of inclusivity and resilience in STEM, making it easier for those following in my footsteps.*

The Unexpected Setback

After six impactful years as the Divisional Head of IT, during which I spearheaded transformative changes and saved the institution millions of dollars, circumstances took an unexpected turn. Despite my unwavering dedication, measurable successes, and undeniable progress, internal pressures and politics nudged me toward an early exit. This decision not only shocked my peers and subordinates but also garnered disbelief from the industry, which had come to respect my benchmark of efficiency and innovation.

Ironically, despite breaking barriers and advocating for diversity, I faced opposition for my dedication to integrity, accountability, and transparency. I had ruffled the feathers of powerful men and refused to conform to a "queen bee" syndrome. Consequently, I was pressured to step down, with offers of significant compensation to facilitate my departure.

This setback served as a harsh reminder of the systemic issues that professionals, particularly those from diverse backgrounds, often grapple with in STEM industry. It underscored the importance of not only breaking glass ceilings but also ensuring they remain shattered.

In the face of adversity, my focus shifted to the legacy I was leaving behind—a foundation of efficiency, innovation, transparency and fiscal prudence. I had created a roadmap for others to follow, demonstrating that even in the face of institutional politics, one's true worth, integrity and impact could not be erased.

Lessons Learned and Future Aspirations

The journey of "figuring things out" resembled a winding road, complete with stumbles, wrong turns, but also breathtaking viewpoints. One significant lesson emerged from this odyssey: mistakes were not setbacks but valuable teachers. Each hurdle made me stronger, wiser, and more resilient. I learned that challenges in STEM could be overcome with persistence, dedication, grit, resilience, perseverance, and bravery.

Looking ahead, my vision is clear. I envision a STEM world where leaders emerge from every background, continent, and walk of life, regardless of their intersectional identities. I envision a world where the next generation encounters more open doors than barriers, where aspiring young girls confidently proclaim, "I can be a leader here," where diversity is the norm, and leadership is accessible to all, regardless of background. My dream is to pave a smoother path for those who follow.

Mentorship and Empowerment

Stepping down from my role as the Divisional Head of IT prompted me to reflect on my journey and consider how I could prevent others, especially those with similar intersectional identities, from facing similar challenges. Although I had a few trusted colleagues as a support system and mentors, I lacked a formal mentor or sponsor especially ones that look like me.

My journey as an educator, mentor, patron and empowerment advocate began. I embarked on a mission to empower others through various platforms, including teaching at prominent institutions, speaking at conferences, facilitating workshops, and mentoring individuals through organizations like WiCyS (Women in Cyber Security), Girls Up, Women Entrepreneurs, and Women in Project Management.

I learned that mentorship is like protecting a young plant at the mountain's base, offering guidance and support to help it grow strong and withstand tough conditions. It involves experienced individuals helping newcomers avoid mistakes, overcome challenges, and reach their goals.

Empowering women, particularly in leadership roles, was just like providing them with the tools to construct their own success. By uplifting other women, I created a chain reaction that not only propelled more women to leadership positions but also enriched industries with diverse perspectives and talents.

Conclusion

Reflecting on my journey, I am proud of the obstacles I've overcome and the growth I've experienced. My identity and experiences have enriched my perspective, allowing me to contribute unique insights to the world of STEM.

The message in my story is clear: you're not alone on this journey, and with resilience and determination, you can shape the future of STEM. My experience serves as a call to action for the STEM community and its allies to dismantle the persistent systemic and invisible barriers, ensuring that everyone's skills and talents are valued. Our goal extends beyond adding diverse faces; we aim to celebrate diverse voices, creating a world where every individual can lead regardless of their background. By sharing my story, I hope to inspire others in STEM to advocate for inclusivity and equal opportunity. Let's continue to push for change, dismantle barriers, and celebrate every voice in STEM. The future of STEM should be as vibrant and diverse as the world around us, and my journey invites everyone to help build a world that values diversity and embraces every potential leader.

A New Beginning – The Juanita 2.0...

Retirement marked a new chapter in my life, leading to founder and president Dynamic Solutions International Inc. (DSI). This role presented an opportunity to chart a course based on personal values. The transition brought about remarkable changes—joint pains and headaches became a thing of the past, and newfound energy and contacts revitalized my life.

Through my role as president of DSI, I will continue to inspire, mentor, and empower others. My mission is to make it easier for women to ascend to leadership roles, particularly those with intersectional identities. The journey continues and, together, we can build a world that values diversity, inclusivity, and the potential of every voice. This is my chance to continue inspiring and empowering others, especially women in leadership and advocating for a world that values every individual's contribution.

With every challenge, I discovered a new strength, and with every step, I paved the way for others. My journey as a woman of color, immigrant, and mother in STEM has not only shattered barriers but has shown that leadership isn't about reaching the top alone—it's about lifting others as we rise. ~Juanita DeSouza-Huletey

Juanita DeSouza-Huletey

Juanita DeSouza-Huletey is a distinguished leader in STEM and a catalyst for inclusive change with over 30 years of expertise in Enterprise Information Technology and Project Management. As the Founder, CEO, and Lead Coach of Dynamic Solutions International Inc. (DSI), Juanita has established herself as a trailblazer in both industry and community, known for her approachability, assertiveness, and adventurous spirit. She wears many hats as a leader, speaker, educator, mentor, life coach, and patron, demonstrating a deep commitment to advancing Equity, Diversity, Inclusion, and Belonging (EDIB).

Juanita has made significant contributions to various organizations, notably as the Divisional Head of IT at the Winnipeg Police Service, where her innovative solutions led to substantial cost savings. Her academic involvement includes teaching post-graduate programs at the University of Winnipeg and Red River College, conducting workshops and presenting at international conferences. Juanita holds a master's degree from the University of Manitoba, where she was also a PhD candidate, and she holds several professional designations such as PMP and Advanced Law Enforcement Planner.

Her commitment to community service is reflected in her roles on several executive and advisory boards, including the Advisory Council on Economic Immigration and Settlement and the Board of Governors for the University of Manitoba. Passionately empowering immigrants, women, and girls in STEM, Juanita strives to create environments that celebrate diversity and promote equal opportunities.

Residing in Winnipeg, Manitoba, with her husband Emmanuel and their three adult sons, Juanita embodies the Canadian dream. Her leadership continues to inspire and mentor others, making a meaningful difference in the lives of those within and beyond the STEM fields.

Connect with Juanita at www.dsiconnect.ca and on LinkedIn at https://www.linkedin.com/in/juanitadesouzahuletey.

CHAPTER 13

Leadership Is Who You Are... Management Is What You Do...

Karen Bass

This chapter is dedicated to my family who has always stood by me through thick and thin...thank you.

Leaders are first-born?

Firstborns tend to possess psychological characteristics related to leadership, including responsibility, creativity, obedience, and dominance.

Austrian psychotherapist Alfred Adler

As the oldest of four children, to my loving and devoted parents, it was always assumed that I was the defacto leader behind them. I didn't ask and I made up many of my own rules, processes, and procedures on how to manage my younger brothers and sisters… I was most definitely their leader, and I was managing the heck out of them … like a dictator most times … you know, "do what I say" and "because I'm the oldest." They soon formed a union and enlisted support from the parents which quickly put an end to my limited rule. While birth order is interesting, it's also an uncontrollable factor, so not the strongest indicator for leadership.

Figure 1.0 My first Org Chart- Birth Order

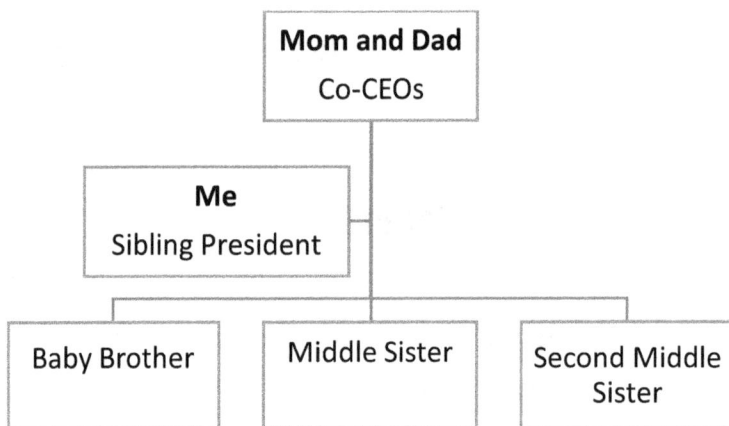

```
                    ┌─────────────────┐
                    │  Mom and Dad    │
                    │    Co-CEOs      │
                    └─────────────────┘
          ┌─────────────────┐
          │      Me         │
          │ Sibling President│
          └─────────────────┘
┌──────────────┐  ┌──────────────┐  ┌──────────────┐
│ Baby Brother │  │ Middle Sister│  │ Second Middle│
│              │  │              │  │    Sister    │
└──────────────┘  └──────────────┘  └──────────────┘
```

Learning to lead through sports

Sport and team activities give opportunities for participants to come up with a game plan and strategies to win. Leaders always have the ultimate goal in mind.
~OC Tanner

Growing up in the south, I participated in several sports including gymnastics, cross country, cheerleading, and soccer, so I got to play on teams while also competing as an individual. I was a fair gymnast for those times and this sport taught me a lot about life from managing through discrimination, building grit, and developing relenting determination. Back in the early '80s in Marietta, Georgia, there were not many black girls participating in competitive gymnastics, a sport where you were judged on your style and execution with much of it left up to the judges' subjectivity. There were times when I was unfairly marked down and I knew it, but I learned early that life is not always fair, and it doesn't mean you give up. I believe that experience gave me the grit and fortitude to overcome when things get hard. Grit and determination seldom hurt anyone and often help to provide the building blocks for leadership. Is it possible these were some of the building blocks I needed for a sales career in STEM?

While I consider myself more reserved, I genuinely like people and I enjoy being an active participant in whatever I am doing. I found that over the years I was a much better coach than I was a gymnast or a cheerleader. I did cheer in college at Georgia State for basketball. While it was fun, I was clearly not the best cheerleader, but it did provide a platform for community exposure and volunteerism. I enjoyed spending time giving back by working at cheer camps for middle and high school girls. I also became a gymnastic coach for over ten years at the gym where I once took lessons. In the same way many parents want more for their children, I believe leaders take what they know and try to make their teams better than what they were. Good leaders are not selfish, and they give of themselves through mentorship or coaching.

Early in my career, I played competitive tennis where I had the opportunity to be the ALTA (Atlanta Lawn and Tennis Association) team manager which meant I collected the membership dues, arranged for team practice, and set the competition line up. It was more of a management job than a leadership role…there were tasks that needed to be executed. As for the tennis team, I saw my role as the manager and our coach as the leader; he was the one teaching and inspiring us to win the match. I was just managing the process of running the team. Both roles were needed but very different.

Fun fact: *EY's research shows that 94% of women executives participated in sports growing up.*

STEM Leadership requires commitment

As I grew in my career as a sales leader, I began to really lean in on who I am as a person to frame my leadership style inside of the technology field where women remain disproportionately underrepresented. I do consider myself a servant leader because that style embodies much of what I believe is important and how I want to exist in the world. Here are ten key characteristics of servant leadership according to Aída Lopez Gomez in her LinkedIn article, *"What is servant leadership and how can it empower you team?"*

10 Key Characteristics of Servant Leadership

Listening	Empathy
Healing	Self-awareness
Persuasion	Conceptualization
Foresight	Stewardship
Commitment to grow people	Building community

As a STEM sales leader, I made it a point to support Women in Technology organizations whenever possible. My membership was a small part in helping to lift up other women. I also find that senior women leaders in STEM organizations are more willing to share their stories with other women. This helps everyone, including those women leaders who by sharing their stories get to lift up others. Leaders are mentors, sponsors, and champions; the best of them always pay it forward.

At this point in my career, I am not managing others directly but lead by supporting STEM organizations that promote women and girls. This is how I can continue to collectively lift others through the support of women in technology organizations and other non-profits, like Girl Scouts and Girls Who Code, that are focused on lifting up young people.

People cannot be what they cannot see and that means that I must get out there so that young people who identify with me can see that there is a place for them in technology. This kind of engagement requires active participation in what matters and a desire to choose to be a leader. Anyone can be a manager and follow a process and that isn't always a bad thing. At times, leaders need to lean in on various leadership styles, but without a doubt, everyone has a dominant style.

There are many tools that can help you assess your leadership style. I think everyone should know where they fall within the DISC profile, a personal assessment tool. Taking the <u>DISC</u> or any other leadership profile assessment will enable you to have a better understanding of yourself and how others view you. These are all things that will support your ability to be a great leader. Increased self-awareness has many other benefits, but one thing for sure is that leaders are self-aware.

Democratic Leadership. ...
Autocratic Leadership. ...
Laissez-Faire Leadership. ...
Transformational Leadership. ...
Transactional Leadership. ...
Bureaucratic Leadership. ...
Servant Leadership. ...

Whenever I started a new leadership role where I was responsible for managing other people, I always did an abbreviated DISC or BIRD profile with my team. It's a fun icebreaker that enables me to quickly learn about the team and in many cases allows them to gain some insight into me.

*Leaders are learners...they are curious and always willing to accept new challenges...*an unknown KSU MBA

It could be said that some people are more natural at leading because they have been put in positions to lead early in life; however, leadership can be learned. I went back to school and received an MBA focused on organizational leadership. I used to wonder if and how I used what I learned but then I realized that I do use it every day. There are many different leadership styles and models that can be explored or studied. I believe that you must understand the difference between management and leadership when you place yourself in charge of other people. Women in particular need to be very clear on this dichotomy of leadership styles and management models. To be clear, I am not your work wife or your mother...oh yes, this is a thing.

I consider myself a collaborative leader, but I do like to have a management model to follow. It is important to be able to manage a process; for example, when the sales leader needs to follow a process for forecasting or performance management. Where many people lose the command of their teams is when they make the management process their leadership style. This can be tricky at times because there are occasions when leadership requires a more tactical approach to avoid risk or to move a process along. The management might take precedence for that given cycle. Effective leaders let it be known that this is going to be tough cycle; however, we will get through it … together. Leaders inspire, motivate, promote, and reward while management measures, records, and documents a process. People need empathy, encouragement, and support. Many management processes are becoming more automated, less human, and not in need of emotion. As leaders in STEM fields, this is even more important to grasp given the heightened concerns around artificial intelligence and machine learning. People matter, and leaders are required.

Leaders play chess not checkers

There is a certain amount of mental patience needed to play chess versus playing checkers. You must be willing to learn who the players are in your organization, you must be willing to research what makes them tick and most importantly, you must be willing to listen to what they reveal about themselves. Once you get an understanding of who they are, you then need to understand who is the king, knight, bishop, pawn, rook, and that ever so clever queen…ha. Don't assume that it's you! Once you have diagramed your working chess board, and you believe you know who's who, then good leaders wait and listen. This is how you learn the politics of your organization and understand who your people are so that you can better understand how to help them, yourself, and support the organization. The goal after all is to get the competition's king.

This is all easy to say now after years of playing spades, tonk, checkers, and anything other than chess. What I mean is I was not strategic, and I thought that my loyalty, work ethic and talent would be enough to be a great leader. Don't get me wrong, you need all those things, but you also need to add in the other elements such as listening to others, researching how things really get done, and assessing your people in the work environment. This is especially true when working in technology driven organizations because things are constantly changing. There is an endless diversity of players in tech organizations; they are vast and usually include a network of external partners that become an extension of people you need to engage, support, and manage.

Earlier in my career, I didn't take the time to assess this situation. *I was playing checkers while the real game around me was chess.* This lack of understanding was evident when I was a first line manager in a global technology company many years ago and found myself being asked to take a lateral role because my local VP said that changes were on the horizon. Soon after that conversation, I met with the Senior VP and shared with him that I was asked to take a lateral position. He immediately said that I shouldn't take a step backwards or sideways and that I should reject the offer from my local VP...so I did. What I didn't realize was that the local VP was really the king! I soon found myself a part of the group that was on the list to be IRIF (involuntary reduction in force). The crazy thing is that I trained the manager that took the role I turned down. This bump in the road could have been avoided if I had a better understanding of who the players were on the chess board. They say experience is the best teacher, but I hope you don't have to make this same mistake.

Leaders lift up others

There is a concept known as servant leadership which was first coined by Robert K. Greenleaf in 1970 in the essay "The Servant as Leader." It literally means that you are focused on your team more than yourself or even the organization. I used to tell my teams that I would take a bullet for them. I prided myself on blocking and tackling the internal minutia so they could just sell and focus on their role in the organization. I was always looking to train my replacement because I had higher ambitions for myself.

As the oldest of four, I was used to putting others before myself. Women are often born servant leaders due to some biology or sense of nurturing that may come from being a mom and always putting everyone in the home ahead of themselves. Consider yourself lucky if you ever get to work for a true servant leader…you will be able to know that your manager has your back. While I was feeling good about myself being a servant leader, I couldn't understand why I felt I was still falling short of being fully effective. As always, the devil is in the details. I forgot that I needed to place my own oxygen mask on securely first before I could save anyone else.

Successful servant leadership is driven from the top down in organizations. You can be a servant leader first line sales manager inside an organization that is metric driven, micro-managing, need-to-know basis driven and you may find yourself leaning into the wind at every turn. I found myself in this situation a few times and decided that I would do my best to support my immediate teams by helping them understand how to advance and actually supporting them. I was fortunate enough to have Xerox Corporation pay for a good bit of my MBA. When I found myself as a new manager, I was able to support another employee to get his MBA. If you wanted the company to participate in paying, the leadership team had to sign off. I was glad to be able to pay it forward so quickly. I felt I was able to lift someone up in that moment. It also made me feel like a successful servant leader when I was able to help one of my employees get a promotion to another team.

Leaders continue to evolve

Leadership in STEM, or otherwise, is not passive; it requires conscience and active participation. You will have the highs you remember from playing team sports like soccer or basketball. You will experience hard and sometimes uncomfortable conversations. But in each case, you will grow as a leader, and you will learn and build upon those experiences. Two things can be true about being a great leader and manager. Strong management strives for efficiencies and continuous improvements which are key components to leadership. I think it would be difficult to be a great leader and a poor manager; however, I do believe you can be a strong manager and a poor leader. I really want to stress that leadership is who you are and will become. Leadership is the personal journey that you travel throughout your life and your career. Management is more of what you do. If you are a leader of people, own it!

Call yourself a leader.

Say it now, "I am a (fill in the blank) LEADER!"

Karen Bass

Karen Bass is an Agile Sales Leader with over 30 years of experience in Fortune 1000 companies such as Oracle, Xerox, Dell, ADP, Konica Minolta, and Toshiba as both an individual contributor and people manager. Karen has an MBA and a Lean Six Sigma Black Belt designation and has also worked as a consultant with her own company, Bass Business Solutions.

She balances work with spending time and supporting STEM organizations like Women in Technology Tennessee, Texas, and Georgia; she has been an active contributor in organizations such as Girl Scouts of Greater Atlanta Marketing Committee, where she chaired their first charity golf tournament, and the Atlanta Beltline TADAC (tax allocation district advisory committee) board where she served as Secretary.

She and her husband Anthony enjoy golf, spending time with family and fishing. They are happy empty nesters living in West Midtown Atlanta.

Connect with Karen on LinkedIn @KarenBass.

CHAPTER 14

Turning Trauma into Triumph: A Journey of Self-Discovery and Success

Karina E. Passi

Thank you to the many parent-like figures that came and went in my life, and to my therapists.

So this chapter will be designed with more action items to get you on your way.

I'm Karina Passi, and I'm a child of a single father who had drug and alcohol issues, and an absent and a schizophrenic mother. I'm a Caucasian female with siblings and cousins of other mixed clines. There was a lot of neglect, abuse (emotional and physical) and much more. I'm glossing over this because, "Where I come from doesn't determine where I went and where I'm going in my life." I should've been a statistic, because I scored high on the ACE test. I'm writing to you as a US Air Force veteran, entrepreneur, essentially healed (with years of therapy), a dancer, a student of life at this point, and I still have more to accomplish.

Growing up lost, I had to orient myself, and figure out my identity. I was punished incessantly for just being, and that crushed my soul, and created a lot of shame within me. School, specifically math, was the only thing that made sense to me growing up, and it gave me peace in life. Realizing that was one of my strengths, I held onto it for dear life, as that was a main source of self-esteem. I was so fragile then, that not doing well on a test was emotionally difficult to process. I was able to get my angst out with sports, and that ended up being a secondary source of esteem. Growing up while externally sourcing my self-esteem would change in my late 20s. During high school, I read the book by Cynthia Kersey, *Unstoppable*, and created the road map to my dreams, but first I had to create, find, heal and grow myself.

When you're starting off, make it a priority to do that. Take a legit Myer Briggs personality test, a Strong Interest Inventory test from O*net, or whatever is the new "it" thing, as long as it's useful information. Then use that to your advantage. Play into your strengths, and if you can, play into your weaknesses, too. Look into self-actualizing content and take deep dives into the meaning of every single thing about you. Also know, some parts of you will change, and other parts will stay the same. I started as an ENTP (extraversion, intuitive, thinking, perceiving), and after healing my "people-pleasing" and mostly solo-ing through the pandemic, I feel more comfortable as an introvert.

This should go without saying, but whenever you can, take steps to heal. I started therapy as soon as I could in the military. Before then, I was reading self-help books since the age of 16. This was a long journey that I needed to take to be able to succeed. You see, I tried community college, but was struggling being a full-time student, so I could compete in sports and possibly get a scholarship, all the while working full-time, paying rent, and pay my way through school. I was dropping the ball everywhere, except for math. I was self-sabotaging, burning myself out, and I was generally unhappy. I didn't realize it then, since moving from home and moving out of the trauma environment, that poor coping was running my life. All the poor coping mechanisms, such as the people-pleasing and not being able to say no, and more, was running me ragged. I knew this was no way to live, but I didn't know how to get out of this place.

If you're young, take risks (for opportunities), and say yes to safe work experience. What's the worst that can happen, someone says no, and nothing changes? I joined the US Air Force when I was 23 years old because I wanted to go to school for engineering, and I didn't want to take out debilitating loans. Remember how I was feeling stuck earlier? If you're going to join any of the branches, I recommend the Space Force, Air Force, or any branch with science, R&D opportunities.

Gain as much experience as possible, and whenever you find something that you're good at, keep at it. Also, if you see something everyone else is terrible at, get good at that thing, too. You'll be that go-to person for "that" thing. In the Air Force, a lot of the guys were afraid of heights. As a child, I'd play on the roof (yes, you read that right) and in the rafters. I was also in gymnastics. All this to say, I wasn't afraid of heights, but you know who was? My co-workers. So, I delegated myself as the boom lift operator, and I elected for the jobs that were "up there." While I was in the military, I was taking classes for my general ed. and meeting the requirements for an A.A.S. in Avionics Technology. I made sure I had plans. Plan A, plan B, plan C, all the way to plan ZZ. I had nobody to rely on but myself. I was alone. If I failed, I wouldn't be able to eat. If I didn't have rent, I wouldn't have a place to sleep. You get the picture. One time I ran out of gas. One lady gave me $10 USD, and that was enough to get me home. I'll never forget that gesture. Being responsible for so much, for so long, there's pros and cons to it. I should mention that I started working at the age of nine as a receptionist for the family business back when minimum wage was $7 an hour. So, when it was time for me to apply for my first non-family job, it was somewhat easy to get work. Pros = "hella" work experience, phone work, customer service exposure, time and people management, running the point of sales. Mind you, this was pre-computer era. We had colored pens, notes, pagers, we're talking old school. Cons = child labor laws, abuse, and so on.

Get gritty. Gain experiences and do hard stuff. This will boost your self-esteem, and work out your "try" muscle, and your "keep at it" muscle. Unfortunately, being raised in an abusive household, you get used to it, and that could be mistaken for grit. I sure thought it was grit. Abuse actually strengthens your "crap fitting" muscles. Figure out if that's what you're doing. You need to let the crap-fitting skill go, and work on setting boundaries. Boundaries aren't to control or punish others, they're for you! Like Boolean logic, **if** crap A happens, **then** I will do consequences B. No punishing, shaming, just consequences. The BS in military service can make you learn all of these things, as long as you take the opportunity to work out your "psychological muscles." The grit is built into the military experience.

Know your audience, and sometimes it's better to ask for forgiveness than for permission. When I was working with crusties, aka old-timer thinking, they made up their mind about me just because I am a woman. Knowing who I'm working with, I'd do the thing that I know is a great idea (with the ability to revert it if I had to), and just show them the ROI after I did it. I might "move in silence" only showing the results, and they'd never be the wiser. It made my life easier and happier, and they couldn't say no. It was already done.

Then it came time to learn when to say no. After a time, and your experience meter is full, learn when to say "no" to things. Saying "yes" to something you don't want nor need IS saying "no" to something you DO want to do. I said "no" to a toxic work environment in 2019, left, started my consulting business, and said "yes" to being on the leading edge of UAS (Unmanned Aerial/Autonomous Systems) technology!

This was a huge risk. I had no net. The place was so toxic, that leaving asap was better than staying or anything else. The Universe/ God or who/what -ever you worship, had my back. When you move in the light, and speak truth, good things seem to happen. Warriors of the light are protected beings and that's part of our job here. Truth is healing. It's only painful when you choose to be blind and deaf to it. Of course, if you can line up something, please do.

Trust the process. I've only been in love twice. Each breakup destroyed me in a different way. It took me 10 years to allow love in my life after the first. The first one taught me "never move for a boy." This was who I left when I joined the military. Joining the military changed my life forever. Some of the changes were for the better. I went through some pretty rough stuff, and if you're interested in joining, I'd send any gal to the forces with lots of warning and cautions. Just stay away from the men in all sexual and romantic senses. There's a reason the military is in the news for rape, murder, MST, cover-ups and so on. If that guy is really the one, you two can get married when you both get out of the military. I'm going to expand on the "don't move for a boy" and include; don't give up your means of income for one, don't put your dreams on hold for one, and basically don't do anything that he isn't willing to do first.

The latest love taught me to "trust the process." I didn't get it then. He was not working, it was the pandemic, and I couldn't understand it. How could he not stress about this? I was so used to planning, pivoting, executing, that I didn't leave space for the divine. This might be woo-woo or spiritual speak, but some things are indeed out of your control. The faster you accept that, the faster you know that you're the only thing you can control in life. Having a boss blackball you from future work because you practiced your boundaries? Guess what? If that boss is doing this to you, it's more than likely she/he is doing it to others. "The warrior of the light does not back down." They can't wear that mask forever. When enough people report them to HR, at some point legal action has to be taken. If that company runs off of nepotism, though, you're screwed, there's no winning, and you must always leave. But I can sure as heck mention them in a book. They will always not appreciate you, undervalue you, and make you the bad guy. They will always have high turnaround, unhappy workers, and you're no longer there suffering in that environment. That's the process. It's letting things take care of themselves. When you stop focusing on all the other things out of your control, this will force you to focus that energy on yourself and that's a good thing. Look back on all the good you've done. Now imagine if all that energy went into yourself. You're welcome. Glow up, 10X yourself, and never look back. It's just you, your path, and putting one foot in front of the other.

Be the change you want to see. Sometimes I was one of two women on shift in the military, the only woman in my small job training classes. As I struggled through higher level courses in math, engineering, electronics and physics, I was the only female, or one of the very few. I'm now working with UAS on both the commercial side and military side. To start, I'm probably the first cUAS (commercial UAS) maintainer in US history. I am one of a few female avionics repair technicians in military type UAS, in US history, too. I see you gals. I've worked or messaged you gals, too. I'm proud of you for those of you who did it for a while and left, stuck it through, returned after a break, and everything in-between. Some of the people at these places suck. Straight up. But I thank you. Because I saw you gals, I didn't feel alone. Because we texted, or talked, I felt capable again.

I've assembled, repaired, soldered, pushed, carried all kinds of UAVs (Unmanned Aerial/Autonomous Vehicles), aka drones. This career is versatile, fun, and if you like frequent changes, then it's exciting, too. It fits my personality type. My nitch is startup work culture and working in new departments. I quickly die inside if I'm a cog in the machine doing the same thing repeatedly. Working in the UAS space allows me to travel, meet people from all over the world, and be creative and innovative in some way. It allows me to express my feminine side through my fashion, my masculine when I'm training others, and meets all of my needs in my "need triangle." This is the three main things one would need to be able to work in a workplace, such as proper compensation, workplace culture, and let's say you need your input valued. This is different than Maslow's hierarchy of needs, but with similar principles. I encourage you to research this content while you figure yourself out.

I have to keep this short. All the best to you and your endeavors. I hope this helped.

Karina E. Passi

Karina E. Passi is a survivor of childhood abuse and neglect. She is an Air Force veteran. She's now living what most would call a successful life.

Dear reader, thank you for reading part of my story. This is a first for me, and I'm humbled that you found my story useful.

Connect with Karina at www.LeanProcessSolutions.com.

CHAPTER 15

Empowered to Lead: How Women in STEM Navigate Challenges and Maximize Impact

Dr. Leigh Holcomb

*To my mom, Shirley Holcomb: Mom, your steadfast encouragement
that I could learn whatever I set my mind to do made all the
difference. Yours is a spirit that shines for all to see with the
courage and love that come from a close
personal relationship with God.*

"Why did I decide to become a Neuroscientist?" It wasn't as if anyone in my family was a scientist. When I started to reflect on the answer to that question, I came to the realization that it was mainly women in my life who opened doors of opportunity that created my STEM career.

To become a woman in STEM, a girl must overcome multiple barriers and biases: the educational system, family expectations, psychological factors (imposter syndrome), as well as experiences in the workplace (leaky pipeline, sticky floor, broken rung, and glass ceiling).

Overcoming the STEM Gender Gap

Data Point: There is a small window of 4 to 5 years to capture a girl's interest in STEM. Girls begin to lose interest in science and math by age 15.

The Importance of Role Models

Having visible role models is one of the keys to keeping girls interested in STEM (Microsoft. Why Europe's girls aren't studying STEM, 2017).

My role model was a Marine Biologist named Mariette Coulter. I found out that I LOVED science when she came to teach at my elementary school.

Mariette Coulter's enthusiasm for the ocean was infectious! I still remember a super-cool field trip we took to collect specimens like hermit crabs and baby fish from among submerged mangrove roots. We were even photographed together for a newspaper article examining a shark specimen. This scientist convinced me that I was bright and capable of becoming a scientist.

Data Point: Girls face unconscious biases with STEM and math learning (Copur-Gencturk et al., 2020)

My future career as a global scientist was almost crushed by my 6th grade math teacher. Being called up to the chalk board to work a math problem in front of our class was the bane of my existence! I stood frozen with the chalk in my hand staring at the math problem. "You'll never learn to do math" Mr. Lefers said as he told me to put down the chalk and go back to my seat.

Someone snickered with laughter as I walked down the aisle to my seat and sat down with my head hung low and tears in my eyes. I brought home my first D on a report card with a sense of shame. I remember feeling so very stupid and worthless as I showed the report card to my mom.

Data Point: Women tend to think they will not be good at a STEM career compared to men.

AAUW reported that beliefs about women's intelligence, stereotypes, self-assessment, the experience in school and work as well as implicit bias can impact success in STEM careers (Hill et al., 2010).

Advocates Matter

My mom made an appointment to talk to my math teacher. "My daughter has NOT been a D student. If she isn't getting the math information, then it has to do with the way that you are teaching it!" My mom was like an Avenging Angel sent from God with a flaming sword of truth that cut through the negativity that Mr. Lefers spread to all of his students. "Well, I don't think it will help but there is a box of materials in the back of the classroom that she can try," was his response.

I quickly pulled my grade up in math without the shame-based teaching shutting down my brain. I later learned that a math-learning disability called dyscalculia runs in my family and that my mom had been discouraged from studying chemistry because of her math struggles. I required math tutoring in high school and college but still managed to complete a PhD.

Career Conversations

My first real career conversation occurred when my dad asked the all-important question, "What are you going to do with your Bachelor's in Biology?" "I don't know. I guess I will go to graduate school," was my answer. After earning a MS and a PhD, I began looking for a post-doctoral position in Academia.

Good Mentorship

I joined a women's mentoring program through the Society for Neuroscience (SFN). My mentor's advice was, "Take every job interview offered if it is remotely interesting. If nothing else, it will be great practice and that's one more person that knows you."

I had an interview through the SFN Job Fair. Reviewing the job ad requirements, I started to disqualify myself from even applying as many women do. My self-assessment said I was not a 100% fit for the role. I went ahead with the interview with my mentor's advice in mind. I'm so glad I listened to her! By the end of the conference, I had a post-doc job offer from the group at Texas A&M!

Filling the STEM Work Pipeline

Implicit gender biases exist that portray women as less capable, skilled or have less expertise in STEM (Hill et al., 2021). Women outnumber men in earning psychology and biology undergraduate degrees with fewer women entering engineering and computer science as majors.

Women earning bachelor's degrees in STEM

Psychology	78%	Physical Sciences	40%
Biology	62%	Geosciences	39%
Social Science	61%	Engineering	22%
Mathematics	42%	Computer Science	19%

(Data from Seneviratne, P., 2022)

Men rate other men as more knowledgeable than women in life sciences and physical sciences. Women also rated men as more knowledgeable in physical sciences but not in life sciences where women are more prevalent. Data from Bloodhart et al, (2020) showed that women earn more A or A+ in college science courses and have higher GPAs than men. This bias has implications for retention of women in STEM careers. The "Leaky Pipeline" refers to the loss of women from the STEM workforce. Data from 1.2 STEM college graduates in the US indicates that leaks can occur before college due to lack of STEM readiness as well as in college where women are less likely to choose and complete a STEM major. STEM-readiness accounts for 8% of the gender gap in STEM. Also contributing to the gender gap in STEM is a 16% difference between males and females graduating with a STEM degree (Speer, 2023). A rapidly decreasing number of women go on to earn advanced degrees or into the STEM Workforce.

The Leaky Pipeline in STEM

Women in STEM	BS 50%	MS 44.3%	PhD 41%	STEM Workforce 29%
Minority Women	BS 1.3%	MS 12.4%	PhD 6.8%	STEM Workforce 4.8%

(NSF NCSES, 2019)

In 2021, the STEM workforce was 65% male and 35% female (NCSES, 2023). Mid-career is when attrition of women from the STEM workforce is most prevalent (Hill et al, 2021). By ages 30 and 45, the persistence of women in the STEM workforce was less than for men (Speer, 2023). Women working in STEM were nearly 2xs more likely to say yes when asked if they are thinking about leaving the workforce compared to women working in other industries. Reasons given for wanting to leave the workforce were: Stress and Burnout (32%), Others getting promoted ahead of them (29%), Lack of purposeful/meaningful work (25%) and lack of diversity (20%) (MetLife 2022).

The "Sticky Floor" phenomenon is prevalent in STEM where women are prevented from taking the first step to career advancement. Women with a STEM PhD are more likely to be in a junior faculty role than a man (Di Fabio et al, 2008). Women still earn 82% of what men do. Women with STEM roles in Academia have a larger gender-related pay gap compared to women who work in industry (Ding et al., 2021).

A deeper dive into the demographics of the life science workforce in Massachusetts showed the leaky pipeline. Women are 66% of life science graduates but only 52% of life science workforce in Massachusetts are women (14% loss) (MassBio, 2022). The "Broken Rung" and "Glass Ceiling" effect can be seen as women advance in their careers into higher levels of management. Women are still excluded from the highest paying decision roles in STEM (Tufts University, 2023). Executive management teams are 46% female, increased 9% from 37% in 2021. The presence of women on boards of directors went up from 24% in 2021 to 33% in 2023. Nearly 25% of the life science workforce is over 55+ years old, highlighting the problems of excluding women from progressing into leadership where the upcoming retirement of knowledgeable workers in the Pharma collides with the dwindling pipeline of incoming talent.

Women in Life Sciences

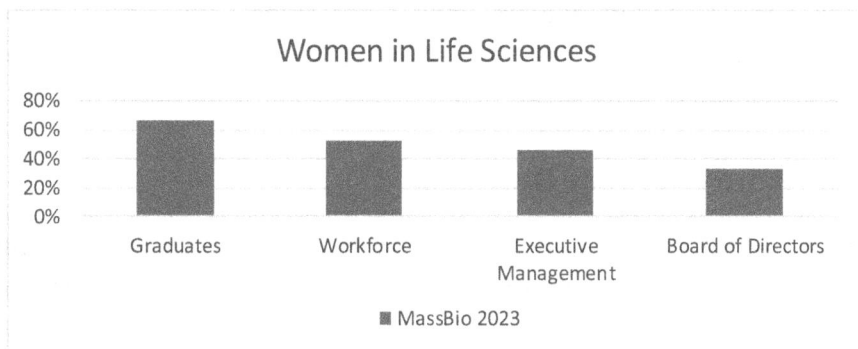

Tap the Power of Professional Organizations

Three of my favorite professional organizations for Women are the Healthcare Businesswoman's Association (HBA hbanet.org), Women in Bio (WIB womeninbio.org), and Women, Life and Science (WLS wlscience.com).

HBA began in 1997 and now has over 17,000 members in 75+ countries working in healthcare and life sciences. HBA has a commitment to achieving gender parity in leadership positions, **providing equitable practices** that enable organizations to realize the full potential of women, and **facilitating career and business connections** to accelerate advancement.

WIB's primary focus is to provide women-to-women mentorship and leadership support through all stages of career development: Bench to Boardroom, Academia to Industry, First Job to Last, and Idea to Entrepreneur. WIB has 13 chapters in North America. One of my favorite WIB events is the Speed Networking sessions, where women at all stages of their careers can connect.

WLS started as a podcast by Cecilia Zapata-Harms (https://womenlifeandscience.buzzsprout.com) and grew into the Science

Cafe network forum and an annual conference elevating women entrepreneurs, leaders, innovators, and educators especially in the science industry.

Career Advice from Dr. Leigh

Through my company, Career Catalyst Edge, I provide career coaching for the life science workforce to help my clients effectively design career paths that align individual values and life aspirations.

Here's some advice that may make your STEM career transitions easier:

Be a Lifelong Learner

After we finish our formal education, we need to be intentional about setting aside time, money and energy to upskill routinely. This effort can be as simple as picking out a great podcast, reading a book that a colleague recommends, or following top voices on LinkedIn in your field.

Become Visible

I had a work colleague that I considered a mentor say to me, "Leigh, I don't think that you ever got the credit you deserve for all the work that you do." That was a thought-provoking statement for me! How visible are you? Do the people you work with understand your strengths and how best to work with you? Think about your personal brand. What do you want people to say about you to others? Learn to use your LinkedIn profile to stand out from the crowd.

Conduct a Periodic Career Check-up

Don't wait for your employer to plan your professional development. No one will care more about your career than you! If you could redesign your career, what would you change about your current situation? Think what you would keep, eliminate or alter (and how would you alter this item).

Save Money for Career Transitions

Make sure you have a nest egg set specifically to give you the option of changing jobs. Layoffs occur frequently in pharma and biotech. The average job search often takes >6 months. You may want to hire a career coach or obtain a new certification, both of which will take money.

Learn How to Conduct an Effective Job Search

Plopping your resume into a company career portal to apply for a job gives you a 2-3% chance of being hired. Every job posted online gets over 250 applications. Resumes typically get 6-7 seconds of the recruiter or HR person's attention during the screening phase.

Connect with recruiters that have expertise in your desired career area. Not every job offering will be posted online. Biotech and life science companies may hire a recruiter to fill complex or time-sensitive searches. One thing that I learned as a recruiter was if a candidate had already applied through a company portal, that I would not be able to represent the candidate. Another reason not to "Pray and Spray" and apply to every job online.

The best way to be considered a top candidate for a role is to get a current employee of the company to provide an internal referral for you. Referred candidates are preferred because they tend to be a better culture fit, stay longer, and are a faster hiring process than selecting candidates from job board applicants.

Be Generous

Excellent career opportunities at pharma/biotech/medical device companies may come your way when you least expect it. If you decide that it's not the right fit for you, spend a few moments thinking about who might want the info. You can be a blessing every day. Somebody's praying for that opportunity that you just passed on.

Pay It Forward Focus

Lift up those women that are younger in their career and pass along your social and knowledge capital. Become a sponsor!

"Sponsorship is much more proactive: 'Let me put my own reputation on the line' to help someone else in their career and amplify, lift them up to that next level in their career." (LaHucik K., 2022).

Champion Neurodiversity in STEM

Neurodivergent women in STEM face unique challenges. Neurodivergent individuals often excel in divergent thinking, creativity, risk-taking or spatial visualization skills. 15-20% of the world's population is neurodivergent. 80% of Autistic women go undiagnosed by age 18 and the average age range for a woman to be diagnosed with ADHD is early 30s-40s. Neurodivergent girls often go undiagnosed until much later in life than boys, missing the critical window of opportunity to engage the girls' interest in STEM. Neurodivergent graduate students in STEM programs face challenges such as internalization of neurotypical norms, self-silencing/masking to make it through graduate school, and neurodivergent burnout due to overwork (Syharat et al, 2023).

As a neurodivergent woman with late-diagnosed ADHD, I have reevaluated my own personal definition of success in STEM. One thing stands clear: the path to true success is about more than technical skills.

It's about building resilience and unwavering belief in yourself as well as the courage to lead yourself, your peers, and the next generation. Success in STEM requires that you find your voice, challenge biases, and build a career that aligns with who you are at your core.

If you're ready to break through barriers, to champion your strengths, and to write a career story that reflects your unique brilliance, I'm here to walk alongside you every step of the way.

Working together, let's cultivate the skills, confidence, and purpose that will empower you not only to succeed in STEM, but to lead with impact and inspire change for those who follow. Let's create a future in STEM that welcomes all that you are as a woman!

References

- Bloodhart B, et al. Outperforming yet undervalued: Undergraduate women in STEM. PLoS One. (2020) Jun 25;15(6):e0234685.

- Copur-Gencturk, Y., Cimpian, J. R., Lubienski, S. T., & Thacker, I. Teachers' Bias Against the Mathematical Ability of Female, Black, and Hispanic Students. Educational Researcher (2020) 49:30-43. https://doi.org/10.3102/0013189X19890577

- Ding, W.W. et al. Trends in gender pay gaps of scientists and engineers in academia and industry. Nat Biotechnol. (2021) 39:1019–1024. https://doi.org/10.1038/s41587-021-01008-0

- Hill, C. et al. Why so few: Women in Science, Technology, Engineering and Mathematics AAUW. (2021) https://www.aauw.org/app/uploads/2020/03/why-so-few-research.pdf.

- LaHucik, K. 'Biotech Sisterhood': 25 female CEOs find new ways to 'sponsor' the next generation at Arizona retreat. (2022).

https://www.fiercebiotech.com/biotech/biotech-sisterhood-25-industry-ceos-convene-sponsor-support-women-c-suite

- MassBio 's 2nd Bi-Annual State of Racial, Ethnic, & Gender Diversity Report: The Progression of Equity and Inclusion in the Massachusetts Biopharmaceutical Industry (2023) https://www.massbio.org/wp-content/uploads/2023/11/2023-DEI-Report.pdf

- MetLife 2022 TTX Survey on Women and STEM. https://custom.cvent.com/97556B35C697414396BE492ACA592CA1/files/41d9ea5cebb14f27acf7396c002d3498.pdf

- Microsoft. Why Europe's girls aren't studying STEM. (2017) https://news.microsoft.com/uploads/2017/03/ms_stem_white-paper.pdf

- National Science Foundation National Center for Science and Engineering Statistics. Women, Minorities, and Persons with Disabilities in Science and Engineering (2019). https://ncses.nsf.gov/pubs/nsf19304/data

- National Center for Science and Engineering Statistics (NCSES). Diversity and STEM: Women, Minorities, and Persons with Disabilities Special Report NSF 23-315. Alexandria, VA: National Science Foundation. (2023). https://ncses.nsf.gov/wmpd

- Nicola, T.P. STEM Gender Gaps Significant Among Gen Z. (2023) https://news.gallup.com/opinion/gallup/544772/stem-gender-gaps-significant-among-gen.aspx

- Seneviratne, P. Are women reaching parity with men in STEM? Econofact (2022). https://econofact.org/are-women-reaching-parity-with-men-in-stem

- Speer, JD. Bye bye Ms. American Sci: Women and the leaky STEM pipeline. (2023) Econ. Edu. Rev. 93: 102371 https://doi.org/10.1016/j.econedurev.2023.102371

- Syharat, CM et al. Experiences of neurodivergent students in graduate STEM programs. Front. Psychol. (2023) 14:1149068.

- The Neurodiversity Podcast with Emily Kircher-Morris Episode 220: Unlocking the power of neurodiversity in STEM. https://neurodiversitypodcast.com/home/2024/4/12/episode-220-unlocking-the-power-of-neurodiversity-in-stem

- Twaronite, K. We know very little about neurodivergent women-and they may be entirely overlooked at work. (2024). https://fortune.com/2024/05/20/neurodivergent-women-work-health-careers-leadership/

Dr. Leigh Holcomb

Dr. Leigh Holcomb is the CEO of Career Catalyst Edge. Trained as a Neuroscientist, she uses her decade of Pharmaceutical Industry experience plus iPEC coach training to support women in Pharma & Biotech. Dr. Leigh has a unique perspective having been on both sides of the interview table and is passionate about helping others find their dream job and career path.

Dr. Leigh has been a speaker at Life Science conferences and an advocate for the advancement of Women in Science. She is an international best-selling author and speaker on Neurodiversity. Raising awareness of ADHD, Autism, as well as Dyscalculia (a math learning disability) are part of Dr. Leigh's current educational efforts.

She is on the Board of Directors for RIZE Prevention focused on school-age drug prevention, Just Bee, which was founded to create Autism and Neurodiversity friendly communities, and Women, Life and Science, a network forum and a think-tank community elevating women entrepreneurs, leaders, innovators, and educators.

Dr. Leigh is based near Greenville, SC where she shares a multi-generational household with her joy-filled mom and her creative daughter, a future animator. If Dr. Leigh could tell you one thing, it would be "Take care of your brain because it's the only one you've got!"

Connect with Leigh at www.careercatalystedge.com.

CHAPTER 16

The Evolution of Leadership in Cybersecurity Amid Persistent Challenges Facing Women

Maggie Calle

To my lovely children and my niece. May God give you wisdom to choose the career path that will make you happy because when you

---

"Find a job you enjoy doing, you will never have to work a day in your life." ~ Mark Twain

---

It was the spring of 1996 when I was overjoyed with excitement to have been accepted to the computer programing and analysis program at the college level; this was the better option for me since university was too expensive. In college, everyone in my class, including myself, had big dreams to go to Silicon Valley to be a programmer and be like Bill Gates. Yes, you heard right, Bill Gates was someone I looked up to. But before Bill, I looked up to my older brother who was taking an information technology and engineering degree in university back home.

Back home, I also was in the first year of the information technology and engineering degree in university and there was no shortage of women in my program so it was only logical that I would continue on that path here in Canada. However, I quickly realized in college that I was in the path less travelled, and my peers wondered how I ended up in my computer programming class, given the lack of women's representation in the technology field. My peers normally asked me what compelled me to pursue a STEM career. More than 25 years later, I sit here writing about my observations of what I've witnessed in my chosen STEM field—cybersecurity.

Diversity, Equity, and Inclusion (DEI) initiatives have influenced the evolution of leadership in cybersecurity, while persistent trends such as bullying, lack of minority representation, and attitudes continue to impact women's participation in STEM leadership roles and mental health. My observations will be drawn from an anonymous survey, Leadership in STEM, given to 165 technology and cybersecurity professionals in my professional network.

Mentorship, allyship and sponsorship have been critical in the evolution of leadership in cybersecurity even though discrimination and bullying continue to emerge in this field, which impacts mental health. First, diversity and inclusion programs with focus on mentorship are one important component to advance into more senior roles in STEM. 81% of the female vs 90% male respondents attributed their success to having great mentors, sponsors, allies throughout their career, and excellent supervisors who gave them career and training opportunities in an equitable and respectful work environment. On the other hand, 8% of female respondents left a STEM field due to lack of work-life balance or discrimination and bullying, which created a toxic work environment impacting their mental health; none of the male respondents had experienced discrimination, bullying or quit their jobs as a result. Finally, 11% of female vs 10% of male respondents did not consider themselves successful in their profession citing they lacked mentors or sponsors.

These findings are extremely significant because they show progress on the one hand, but also show concerning trends. On the positive side, women are making their contributions in the STEM field, while considering themselves very successful. Additionally, a great majority of women surveyed understood key strategies to be successful in their profession and took advantage of DEI programs. On the negative side, discrimination, bullying and mental health concerns emerged as the main reason for women leaving their chosen STEM career.

Not surprisingly, a recent study by WiCys (2023) found that women feel excluded in many exclusion categories including career and growth (57%) and respect (56%). In addition, the same study found that the source for feeling excluded was their leadership (68%), managers (61%) and peers (52%); this study reveals that DEI programs are yet to make a huge difference in this area.

At the start of the pandemic, I attended an online talk on Diversity, Equity, and Inclusion (DEI) by the founder of the Black Professionals in Tech Network (BPTN) and he advised, "Beware of those who come to you in sheep's clothing but are really vicious wolves." This statement resonated deeply within me at the time, as I had witnessed some abusive leaders, particularly toward minority employees, suddenly portray themselves as DEI advocates and allies. I had observed these so-called advocates being the source of bullying and microaggressions where I attempted to address on different occasions by stepping in respectfully and interrupting the moment or calling in afterward privately with the abusive leader to respectfully discuss how their behavior was hurtful. As a result, I found myself or my peers being targeted with harsher mistreatment and microaggressions. Some of these leaders created a hostile work environment, causing employees significant stress and impacting employees' mental health to the degree that many left the organization.

Following the rise in DEI initiatives after George Floyd's murder, these same toxic leaders emerged as the faces of their organizations, using DEI advocacy to further advance their careers. In summary, while DEI programs have contributed to the progress of women in the workplace, challenges such as discrimination, bullying and microaggressions continue to disproportionately affect women's mental health more in comparison to men.

Second, the efforts to increase minority representation within STEM fields has positively contributed to the evolution of leadership in cybersecurity. With a renewed focus on DEI programs, organizations are creating more opportunities for women to enter the cybersecurity field, thereby increasing the representation of women and other minorities in crucial positions. These leaders can significantly influence the ongoing evolution of leadership in these fields. However, it is important to recognize that decades of inequality cannot be rectified overnight. Despite the increase in women holding leadership positions in cybersecurity, significant challenges remain in further enhancing women's representation in these fields. Young women today have more female role models to inspire them, yet the numbers remain low, with women representing only 20% to 25% of the cybersecurity workforce (ISC, 2024).

Additionally, when asked who inspired them to pursue a STEM career, 57% of female respondents said they were inspired by men, and 38% by women. Similarly, 43% of male respondents were inspired by men, while 30% were inspired by women. These responses are significant, as they highlight the importance of representation. Female respondents, in particular, found inspiration from seeing women in STEM, encouraging them to follow similar career paths. More notably, male respondents were also motivated by strong female leaders to pursue STEM careers. This underscores the progress women have made in the cybersecurity field where their success is being recognized and serving as a source of inspiration for both men and women. In cybersecurity, minority representation is critical. It shapes the perception that anyone, including women, can thrive in this field, which is not exclusively for those with hard technical skills. Cybersecurity encompasses a wide variety of roles, requiring diverse individuals with a diverse set of skills, talents and abilities. This diversity fosters innovation, as solutions are best achieved through a blend of different ideas, approaches and diversity of thought that leads to achieving success in the evolving cybersecurity landscape.

Furthermore, this increased visibility of women leaders also suggests that organizations' DEI programs are working effectively to place visible minorities in prominent roles where they can further influence the evolution of leadership in STEM. This progress is making minority participation in STEM fields a more normalized aspect of the industry. However, while we get to a point of parity, we need to acknowledge that there are several barriers that continue to impact minority representation within cybersecurity. In the Leadership in STEM survey, some participants highlighted the persistent issues such as lack of work-life balance that drove them to leave their careers in STEM. These persistent issues reveal that while we are able to bring women into the cybersecurity fields, we are not always able to retain them due to the lack of work-life balance – these seem to be the basic non-negotiable aspects that need to be improved upon by organizations through their DEI efforts.

While I was concerned about the lack of women representation in cybersecurity, there are other major concerns that surfaced in the survey given that none of the survey respondents identified themselves as members of the LGBTQ2S+ while 55% of respondents identified themselves as women and 45% as men. This might indicate that the LGBTQ2S+ community is not interested in a field where discrimination and bullying are still present; however, more data is needed in order to explore this area or come up with conclusions. In brief, while a lot of progress has been made to bring diversity into the cybersecurity field, we still have a lot more work to do to attract more women and other minorities into this field and maintain strong retention rates.

In the ISC2 report Women in Cybersecurity: Women in the Profession (ISC2, 2024), presents a clear picture of how women's representation in the cybersecurity industry is positively changing as more young professionals keep entering this field. However, the same study also reveals that the first generations of women in cybersecurity, likely those currently in leadership roles and ages 45 and up, are a clear minority, representing only 13% to 15% of the cyber workforce. This specific trend suggests the only way to increase minority participation in the cyber workforce is by attracting young people into the field.

Third, DEI training programs for employees to prevent unconscious bias have made a significant difference in the evolution of leadership in cybersecurity. According to the Catalyst, "Unconscious Bias is an association or attitude about a person or social group that, while not plainly expressed, operates beyond our control and awareness, informs our perceptions, and can influence our decision-making and behavior." (Catalyst, 2024.) Although workplaces are proactively attempting to mitigate unconscious bias, these negative attitudes cannot be totally eradicated and are still prevalent preventing women to get into the cybersecurity field and move up the ranks.

Overall, 88% of respondents to the survey think mentorship is a critical aspect to their success; however, 12% of those surveyed did not cite mentorship as a contributing factor to their success. These results reveal a shift in attitudes in both men and women who are taking proactive steps to advance in their career. However, a significant issue not explored in the survey is the difficulty of obtaining a mentor. Many professionals have experienced the disappointment of being rejected by potential mentors. Anecdotally, numerous professionals have expressed to me the challenges they face in finding someone willing to mentor them. Imagine finding the courage to ask a respected professional to be your mentor, only to be turned down with justifiable reasons such as, "Sorry, I have a very busy schedule," or "I am already mentoring someone." Another common situation is where a senior leader in an organization chooses to mentor a single employee within their department on a permanent basis. This informal but ongoing mentorship arrangement can create perceptions of favoritism, as other employees may see this as a conflict of interest. If mentorship is crucial for professional success, it must be implemented correctly, and we all need to address the shortage of mentors.

DEI programs should step in to establish formal, structured mentorship programs that are short-term. This approach ensures a full rotation of mentors and mentees, providing rich and beneficial experiences for both parties. Formalized mentorship programs can enhance the quality of mentorship and ensure that more professionals have access to this valuable tool for their professional growth. I have personally participated as a mentor in structured mentorship programs such as the Rogers Cybersecure Catalyst Mentorship program in Toronto, Canada, called Emerging Leaders Cyber Initiative (ELCI).

This unique mentorship program should be replicated by organizations interested in establishing formal mentorship programs. The ELCI mentorship program lasts for six months and sets clear objectives upfront. With each mentorship period, I have become a better mentor and have learned a great deal from the mentees, making it a mutually beneficial arrangement. To sum it up, whether it's evolving attitudes towards seeking mentorship, the fear of rejection, or the lack of quality mentorship programs, the absence of effective mentorship can hinder professional success in the cybersecurity fields. Mentorship is vital for the evolution of leadership in STEM, and organizations must step up and institute formal mentorship programs as part of their DEI efforts.

I think women are not adequately represented in DEI research and, because of this, the survey was sent to more women than men in order to get enough voices from women. However, a key objective was to understand what men are doing to get to STEM leadership roles that women are not. Therefore, 43 men and 122 women were targeted for the survey; however, the response rate for men was 70% and for women only 30%; this is not a great response rate for women when it comes to understanding issues women are facing. If this is typically the trend, any advocates or researchers would not have enough data to aid and implement appropriate programs that make the workplace better for women and organizations will keep designing workplaces that attract men in greater proportions. I have myself been in situations where I did not participate in surveys or DEI programs simply because I had no time while balancing work responsibilities and my children. I truly believe that the issue needs to be explored further to understand the reasons behind adequate representation of women in research.

In summary, the emphasis on eliminating unconscious bias and fostering awareness has made the workplace more inclusive and appealing to minority groups, enhancing their sense of belonging.

These initiatives, along with shifts in employee attitudes, benefit not only women but all minority groups. The progress observed over my 25 years in the workforce has contributed positively to the evolution of leadership in cybersecurity. These changes have enabled women to actively engage in decision-making within some of the world's largest organizations, paving the way for other women and minority groups. The investments made in DEI training and related programs yield a strong return on investment for organizations focused on innovation as a diverse workforce can harness a wide range of perspectives, providing a competitive advantage.

In conclusion, while women have made significant strides and now hold high-ranking leadership roles in cybersecurity, there are still many substantial opportunities for improvement in DEI programs, as outlined in this chapter. Many minorities are still underrepresented in the field, indicating a need for continued proactive efforts. Outreach programs have successfully increased women's participation in cybersecurity, with women under 35 now representing 25% of new entrants. However, challenges persist—women in cybersecurity often face discrimination, bullying, microaggressions, limited growth opportunities, lack of respect, and difficulties balancing work and personal life. These factors severely affect their mental health, hinder career progression into leadership roles, and even contribute to attrition.

It is clear that both men and women recognize the importance of mentorship as a strategy for career advancement, yet there is a shortage of mentors in the field. Additionally, both value career and training opportunities within an equitable and respectful work environment as critical to their success.

Ultimately, women in leadership roles within STEM are breaking through the glass ceiling and inspiring not only the next generation of women but also men. This is a powerful testament to the visibility and influence of these female leaders who are paving the way for future generations and helping to reshape the landscape of cybersecurity.

Works Cited

- (2022, May 25). Understanding Unconscious Bias: Ask Catalyst Express. Catalyst. https://www.catalyst.org/research/unconscious-bias-resources/

- (2024, April 25). Women in Cybersecurity: Women in the Profession. ISC2. https://www.isc2.org/Insights/2024/04/Women-in-Cybersecurity-Report-Women-in-the-Profession

- (2023, February). The state of inclusion of women in cybersecurity. WiCyS. https://www.wicys.org/wp-content/uploads/2023/03/Executive-Summary-The-State-of-Inclusion-of-Women-in-Cybersecurity.pdf

Maggie Calle

Maggie Calle is the Chief Information Security Officer at Varicent, with over 25 years of experience in cybersecurity leadership across the financial, insurance, retail, and technology sectors. She is known for successfully implementing cybersecurity and risk management programs that enable the business and drive innovation and digital transformations.

In 2024, Maggie was recognized in the CISOs Top 100 Accelerated list for her impact in the field. She also received the 2023 Trailblazer Award from Ted Rogers School of Management, which honors a distinguished Ted Rogers graduate who has demonstrated exceptional accomplishments in their chosen profession and was inducted into its Alumni Wal of Distinction. Other accolades include being named one of Canada's Top 20 Women in Cybersecurity, Women to Watch, a Global Top Influencer in Cybersecurity, and receiving awards like Varicent's Leadership Vantage, Aviva's CIO and BMO's Best of the Best.

A public speaker, mentor, and co-author of the #1 international bestseller *Women Transforming the Landscape of Science and*

Technology, Maggie holds CISSP, PMP, and Chartered Director certifications. Her academic achievements include a Bachelor of Commerce and IT Management from Toronto Metropolitan University, an MBA in Risk Management from Athabasca University, and masters in project management and computer programming.

Connect with Maggie on LinkedIn at https://www.linkedin.com/in/magalicalle.

CHAPTER 17

Breaking Barriers in Male-Dominated Industries

MaryBeth Wegner

*To my parents, siblings, friends, coaches, teachers/professors,
colleagues, and all other mentors—women and men—in my life.
Thank you for always pushing me to grow, and for supporting me
when I falter. I am forever grateful for each of you.*

When I was asked to write about my career as a woman in STEM, I was flattered, and a little scared. Do I really have anything interesting to say about it? I hope you'll think so by the end of this chapter. My experience may be atypical (but probably not), and I may jump around a bit, but I'm going to try to impart some wisdom and a lot of support for each of you through some anecdotes. I'm not sure about the wisdom, but definitely support. WAIT! I did it again! Back to that reference in a minute.

Let's go waaaaaay back in time.

I have always been a bit of a tomboy. Can you relate? I was drawn to traditionally male activities and was usually the only girl playing football, baseball, kickball and other sports with the guys. They would come to my house to ask if I could play sports during the summer. Back then, I was a girl who knew—not just thought—that I could do and be anything. I dreamed of a career in the NFL as a quarterback or wide receiver. Maybe a running back, because when I was on defense, I could take down any of the boys by myself, but when I had the ball, it would take two or more to take me down. A low center of gravity and strong legs have their advantages. In 6th grade, apparently a switch goes off in a boy's mind and they discover that girls don't actually have cooties, but also maybe they don't want to play sports with them anymore because there might be other things they would prefer to do with girls (yet, still they are terrified of girls).

When I started playing trumpet, I didn't realize that it was an instrument primarily played by boys and men (I'm so glad to see this changing today). After graduating high school, I went to college at Notre Dame, a male-dominated college that had a 70-30 men to women ratio when I started and around 65-35 by the time I graduated. I'm happy to report that according to U.S. News & World Report in 2022, the ratio was 51-49, so it's improving tremendously, but there is still work to do. While at Notre Dame, I stumbled into my Geology major, another male-dominated group, which led me to a career in the oil and gas industry, where not only is it male-dominated, but when I started it was a largely "Good ol' boys" male domination. We'll come back to this in a minute.

Getting back to my "not sure about the wisdom" comment at the beginning of this chapter, I'm curious, can you relate to this? When I was in school, and often in my oil and gas career, I would watch my male colleagues seem to be very confident and just go for it, say what they thought and ask for what they wanted. All the while I was thinking, "I don't know enough to have a suggestion or an opinion. I must need to learn more." Raise your hand if you've ever felt this way? If you said yes, you are not alone. For me it meant getting a master's degree and then starting a Ph.D. I still don't think I know enough. But I have finally started to learn that I do know a lot and I have the ability to learn and figure out stuff as I go. I only just bought into that, really, in the last few years. I hope that you have known that about yourself for a long time, but at least please know it going forward.

When I was first starting in the oil and gas industry as a summer intern, I had the opportunity to go out for an overnight trip in the Gulf of Mexico on a jack-up rig to relog a well. I went with an older male colleague, and we took a helicopter to the ship. I was going to be the only female on board with about 20 men. The captain had to give up his sleeping quarters for me since he had the only single-sleep room. When I got on board, everyone was super polite and I quickly understood that they were pretty much terrified of me. Well, not ME really, so much as offending a woman on board in some way so that I might file a complaint and they could lose their job. Yeeps! I don't envy that kind of pressure! No thank you. Also, me being me, I don't like people walking on eggshells around me because I think I'm pretty down to earth and easy-going, so I broke the ice with them with a slightly, but not terribly, off-color joke. Why did I do that? I did it to let them know that I was cool, but there was a line. I wanted them to relax so that I could, too. And it worked. The rest of the time on the rig was a lot of fun. I even fished off the side of the rig with some of the guys and [grimace] lost one of their brand new lures… We were very high above the water and I didn't feel the nibble. I still feel bad about that, even though he assured me it was fine.

Humor can help in some situations with finding common ground. I think that men are often just scared of women these days, not knowing where our individual "lines" are, what they can and can't say around us. But we also have to stand our ground and make sure we're being heard. We have skills, knowledge, and great ideas, as well as a stick-to-itiveness that is borne from feeling like we need to prove ourselves in our chosen fields over and over and over again. We wouldn't be in these fields if we gave up when facing a little adversity. We are capable, intelligent women. Hear us roar. Or at least listen to what we have to say.

Early on in my career, when I first moved to Houston, I had a lot of older male colleagues call me "darlin'." I know it was partly a Southern thing, but still. No thank you. I had to put the kibosh on that. While in oil and gas companies and software service companies, I often felt invisible after a while. I stopped asserting myself, because when I tried, I was often interrupted. When I put forth ideas and real suggestions based on customer feedback, it was always, "That's not in the five-year plan." End of discussion. Very few people asked for my opinion, so maybe I kept offering it, but I stopped trying to fight for it.

You know, I sometimes wonder where that confident tomboy went. I knew I was smart. I knew I was pretty good at sports if I wanted to be. Maybe when that switch flips for boys, it also switches for girls. Boys get more confident; girls get less confident. We start worrying about what everyone, especially that cute one we have a crush on, thinks about us. At least that seems to be part of my story.

Bringing it back to my jobs. Overall, I've been very lucky in my jobs and at least with the men I have and do work closely with, if not the higher ups or those in older generations. My current team is very respectful of each member's expertise and general intelligence. I've never felt more valued. Sometimes we women on the team still get interrupted by our male colleagues, but they are then usually pretty quick to apologize and—after they finish their thought—bring it back to us. If they don't, I'm getting better at bringing it back to us myself. I am actually asked for my opinion/thoughts/expertise with things and then they really listen. They WANT to learn from me. Honestly, it terrifies me. I've not been in this position before. Clearly, I'm still working on my confidence with my own skills, but I'm definitely getting better.

In fact, I am lucky enough to have a team of intelligent young men and women that I get to mentor and teach now. It's been a long time since I have been in a position to teach, not since I was a graduate student. I love teaching and especially love figuring out how people learn and teach them in that way. I love seeing the light go on after they have been struggling with a concept because I finally figured out how they needed to hear and/or see the information.

Let me back up. When I was in high school, I still thought I was smart. No. I still knew I was. It was largely because teachers told me I was smart. I got straight A's without having to work very hard. My physics teacher tried convincing me to go into physics or engineering. That was going a hard no. I love the "why" of physics, but I really don't care about the "how much." I think it's super cool that a road needs to be banked at X degrees if it curves Y degrees so I can safely drive 50 mph on it without flying off the road. I am so thankful that there are people who want to know and are good at figuring that out. That person is not me. I am not a particular fan of math. I mean, I know math is amazing and very necessary for pretty much everything in the world. But I think I don't care for it much because I have only once been taught math the way that I need to learn it, and that happened too late in my math journey. In high school, I got straight A's in all the math subjects, including Calculus. I think it was because I memorized the formulas. But no one told me why I should care about the area under a curve (now I know this is extremely relevant and important for many, many things). When I got to college and was in Calc 1, I thought it would be fine. It was not.

This is how I learn math. I cannot read it and understand it. I see all those formulas, especially the really long ones (I get $E=mc^2$, but beyond that…) and my head automatically says "blah" and I move to the words again. And I don't get it. However, if a person is at a chalkboard/overhead projector/whiteboard/smartboard and, while talking about the elements of the formula, they write the characters and explain them as they write (throwing in the "why it matters" is especially helpful), I get it. They could even use the exact words in the textbook and write the exact same formula, but if I hear and see it happening, it makes so much more sense. Unfortunately, I did not have a teacher or a TA who taught like that until my differential equations course, which was the fourth of four math classes I had to take for my geology major.

All that to say we need to understand that everyone learns differently. One of my favorite things is figuring out how someone learns and modifying my words and teaching style to their learning style. Are they visual, auditory, abstract or other style learners? Once we both figure it out, seeing that "aha moment" is so worth it! I don't mean to brag, but I think there are a couple/few geologists in the world that wouldn't otherwise have gone into geology because of how we connected in the learning.

In a male-dominated career, how do we as women navigate these workspaces when, unfortunately, we are still underpaid, when some men still dismiss our skills, and possibly they make comments about our appearance where they wouldn't to a male colleague? I am afraid I don't really know. I can say all the usual encouraging things you may have been hearing lately:

- Stand your ground.

- Show confidence (even if you don't feel it) when in a meeting.

- Don't be afraid to speak up.

- If a male colleague interrupts you or another female colleague when sharing ideas, calmly let them know that you (or she) were not finished yet and then complete your thoughts.

- Have patience.

- Perhaps most importantly, find male colleagues who will champion you and all your female colleagues, from CEO to Administrative Assistant.

- Find—and be—a female colleague who does the same.

- Mentor younger women in your company, in your field, and anywhere a young woman needs a solid female role model that looks like her, shares her interests, and/or just cares about her and her future.

I personally also think it's very important to have a sense of humor. Perhaps it's just my bias, but it usually works for me.

We are all intelligent and valuable colleagues. We are strong, but often—unwarrantedly—feel like imposters. Find other women—and men—who will advocate for each other to make sure all team members and contributions are valued, regardless of where they come from—the people and the contributions. We also need to be our own advocates and support our female colleagues. Find young women—young people—to mentor. We've got this. It's what we do.

The real takeaway I'd like to leave you with is this gem, paraphrased from something my high school cross country coach, Mr. Johnson, said about running hills:

"Do not let the hill mentally defeat you. You defeat the hill. Attack it head on. **Do not change your pace. Do not change your breathing.** *You will have to change your* **stride** *running up [shorten] and down [lengthen] the hill, but do not change your pace or your breathing. You attack the hill. It is psychology. You have to mentally defeat the hill."* ~Mr. Johnson

I still use that advice today whenever I have a hill to climb, be that hill physical or mental. When complications arise, I don't change my pace or my breathing, but I do change my "stride" to find a way to defeat the issue. Thank you, Mr. Johnson.

MaryBeth Wegner

MaryBeth Wegner is the Problem Validation Program Coordinator at The IDEA Center at the University of Notre Dame. She and her team focus on Problem Validation and Customer Validation, ensuring that the ideas coming through the Center are solving critical problems for one or more customer segments, and that those customers love the solution. She manages this part of the "Commercialization Engine" which helps to create startup companies or find licensing partners for faculty ideas so they can solve real problems and do good in the world.

MaryBeth joined the IDEA Center in April 2022 from the oil and gas industry where, as a geologist, she first worked for exploration companies before moving to the service sector. In her roles in software Technical Sales Management, she focused on problem discovery, customer needs and wants, and solving problems through a customer-first, empathetic approach that created trust, strong relationships, and repeat customers.

MaryBeth earned her BS in Geology at the University of Notre Dame and her MS in Geology at Brigham Young University. She is passionate about solving problems that matter, mentoring, travel, learning new things (currently experimenting with painting and sailing), family, and good friends.

Connect with MaryBeth at

https://www.linkedin.com/in/marybethwegner.

CHAPTER 18

It's Time to Look in the Mirror: Women's Leadership and the Sciences at the Crossroads of Humanity

Phoebe Barnard

This little contribution to the future of humanity is in thanks to my sister, Sue Barnard Lamdin, and the many women and other students, colleagues, mentees, and friends who have shaped my thinking over the years about the power and agency of women at these crossroads of humanity.

Most of us already see the deep dysfunction of human civilization around the world. It's been an unpalatable set of clashing trends – from retrogressive autocracy, political dysfunction, public health crises and economic hostage-taking, to environmental degradation, biodiversity loss, and the grimly accelerating switch of carbon sinks to super-carbon-emitting sources and continental-scale wildfires.

Many might say—although many people still do not see it—that western civilization is in its advanced death throes. And they would inevitably be right.

Why is this inevitable?

It's sadly very simple: the hard, growth-focused core of western civilization—'progress,' entitlement, extraction, inequity, competition, economic proliferation and sprawling consumption and waste—is eventually incompatible with a civilized future.

We've known this for decades. We just hoped it wasn't really so. But now, with over eight billion humans and a net annual growth of over 75 million, the truth is inescapable.

Navigating civilizational crossroads

Civilizations around the world have brought us so much—not least stability, predictability, structure, security, services, diverse livelihoods, sciences, arts, and structured trade—even if they could also lead to oppressive power, slavery, and autocracy, and reduced wellbeing. They are almost universally lauded as a self-evident metric of human progress. Those of us reading this are so used to civilization's benefits that we cannot easily comprehend an acceptable alternative.

So it would be the greatest travesty of our species not to recognize when it's time to move on, and to reshape our dysfunctional civilization—hopefully peacefully, and not violently.

Societal crossroads have already been navigated before at smaller scales. Every post-colonial nation has asked itself 'what kind of a society do we want from now on, and how do we get there?' And we can learn much from their experiences—especially their experiences of leadership through these crossroads.

But to retain a civilization at all, in the hurtling destabilization of the climate emergency, we must face the demons which have driven this destabilization: Our numbers, both human and livestock. Our appetites. Our wastes. And our mindsets. Can we do that?

A recent paper suggests that we can. It is on the human behaviours at the dark underbelly of ecological overshoot: those evolutionary behaviours that used to be adaptive—like our ability to hunt, acquire and protect resources needed for survival, or our ability to chase off marauding competitors and assassins—which are now proving maladaptive, unhealthy, divisive, even suicidal.

With a crowded and stressed planet of eight billion humans, and with the urgency of mass-scale social change in less than a decade, we can (and perhaps we must) use the same tools of global marketing that manipulated our weaknesses for profit—and switch them to making it irresistible to act for good.

The authors of our paper were mainly scientists and educators. But four of our co-authors—strange and yet likeable bedfellows—are disaffected global marketing strategists. They are breaking from the conventional straightjackets of their industry: those goals to manipulate people for others' profit. Global marketing generally persuades people to buy things they don't need and didn't even know they wanted—and then throw them out to buy the next generation of smartphones, to seem smarter, cooler, more wealthy and higher status than the next person. It has been intentionally constructed that way. And it has thrived in a tacit context of making people feel insecure or unhappy unless they buy (or do) what is marketed. But our co-authors decided they could not live with that, and they wanted to shift the industry towards being a force for good.

A future civilization needs a clean break from the past: reckoning with our failure to balance cleverness with wisdom, the present with the future, and self-interest with the common good.

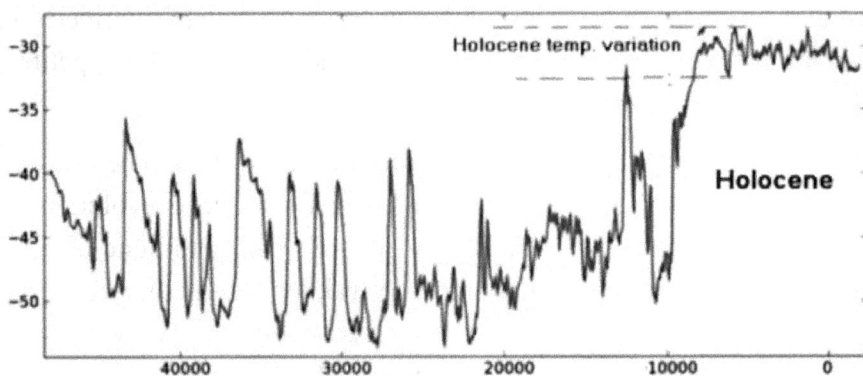

If we can do that, we might just be able to take the good elements of our past with us into a less certain future. It's time to look in the mirror. And science does have a few roles to play.

Science in times of tumult

Like many aspects of human civilization, science doesn't usually prosper in times of rapid change and instability. Over the past 12,000 years of relative climatic tranquility (Fig. 1), known to geologists and climatologists as the Holocene Period, virtually everything associated with the diversification of human endeavor took place. Everything we know and love about human society came into being. The development of agriculture, science, technological innovation, philosophy, arts, cities, temples and cathedrals—all were made possible by this relatively stable, benign climate.

And what this climate stability encouraged was an array of initially small civilizations and societies, increasingly growing and flourishing, on six of the Earth's seven continents: initially in the Middle East and Asia, and later Africa, Europe, Australia and the Americas.

The growth and diversification of civilizations, and the history of the development of science, are both far beyond the scope of this chapter. But it can be reasonably said as an overarching point that the history of science has exhibited four different, overlapping stages: (a) problem-solving practicality, (b) exploration and discovery, (c) interpretation and attempts at unifying theories, (d) technologies for warfare, growth and domination, and much more recently, with the development of the fields of ecology, global change science and climatology, (e) the compilation of disturbing trends into scientists' warnings to humanity.

Women have contributed disproportionately to ecology and global change science—and I myself have labored in these fields for 30 years, and authored and co-authored at least six recent peer-reviewed global scientists' warnings in the last four years.

Women and the sciences of navigating our common future

Women and others with 'feminine' values of care, planning, common-good decision-making and wisdom are at least as likely as – some would say more likely than – many men, at least western men, to take decisions as though people and the planet actually matter. We are perhaps more likely to articulate that everything we know and love is at stake from the effects of humanity's rapid growth.

Can Women's Leadership Drive the Evolution of Our Societies, and of Us, As a Species?

This is partly a rhetorical question. Of course it can. And patently, it must. Whether it will or not, though, depends on the ideological response of society as a whole. We know that social and economic movements often create pendulum swings driven by those whose power is questioned.

Hypatia, the remarkable astronomer and mathematician of ancient Egypt (d. 415 AD), and her academic father, Theon, were among the last scientific intellectuals of ancient Alexandria's enlightened period. For 15 centuries, Hypatia was improbably considered the only female scientist in history. An educated, highly intelligent woman in extremely male-dominated times, Hypatia excelled at teaching Neoplatonist philosophy and mathematics at the Museum of Alexandria. Hypatia's work, like those of modern women in science, focused on how human society could best understand and track its place in the world and history—and her philosophy navigated a tricky centuries-long internecine political and cultural landscape of violence between Christians, Jews and paganists. She was widely popular across Alexandrian society. Despite some modern-day blunt skepticism about Hypatia the woman versus Hypatia the myth, she taught many students practical and theoretical mathematics and astronomy, navigational sciences, and is now credited with writings clarifying Earth's position relative to the sun, despite heliocentrism only being attributed to Copernicus and Galileo over a millennium later.

Hypatia's widespread popularity with Alexandrians of all cultures and faiths was a widely unrecognized political opportunity. She could have continued her work to bring tolerance and the love and value of knowledge to the city. Unfortunately, her popularity was instead seen as a threat to the power of the partisan patriarch, Cyril, and his associate Peter the Lector unleashed a Christian mob of thugs which brutally murdered her in 415, snuffing out the flame of philosophical enquiry and accelerating a descent into a dark age of Christian fanaticism.

Modern society and tackling ecological overshoot

As an evolutionary, behavioural and population ecologist by background, I used to work on the natural and sexual selection of birds. I was fascinated by how evolutionary principles still clearly applied to humans, and thus how we could increasingly apply these fields to the urgent challenges of global change science, and how human behaviour change might happen in order to shift humanity away from the abyss of ecological overshoot.

I remember it being said in my youth that women were more involved in ecology as it was a "softer" science than physics, chemistry or mathematics. Of course, this was seldom true—quantitative sciences across the board are all seen as "hard" and qualitative or categorical sciences are often regarded as softer. Ecology can be a highly quantitative science. And the invisible (largely male) mindset that equates female participation with the soft sciences persists disturbingly today.

In my own case, I worked on biodiversity loss, climate change, land use change, biological invasions of alien species, and environmental observation systems, and was privileged to collaborate with women and men whose quantitative intellects and incredible focus greatly strengthened our international research programmes on species climate vulnerability, endemicity, palaeoclimate and land use change.

But I had always felt that it wasn't about the hardness or softness of ecology that attracted women, but about its urgency for solving real-world problems of our planet, our climate, and our society.

Women are recognized pretty widely—though still somewhat controversially—as statistically highly inclined towards futures thinking, collective well-being, strategic planning, legacy benefits, and moral-compass-based decisionmaking, e.g. motivation towards the future. And while the statistics of conventional academic or corporate metrics are slim and hard to come by, the stories of women themselves frequently emphasize a strong concern for family, planet, climate, sustainability and our common future. There is still much argument about the reasons for this, except in leadership circles where statistically significant discrepancies between men's and women's personality traits are clear.

A groundbreaking leadership programme for women in science, technology, engineering, maths and medicine, with which I am an 'elder-mentor' and inaugural participant, Homeward Bound grew out of a literal dream of its co-founder, the Australian entrepreneur and women's ethical leadership expert Fabian Dattner, of traveling to Antarctica on a boat with women scientists. She woke up with the determined belief that one of the fastest ways to create a more sustainable global society was to increase women's roles: to increase the numbers, strategic skills and visibility of women at the leadership tables of the world. Homeward Bound is now in its eighth year of creating a diverse global leadership community of 10,000 women with a STEMM background, so far with over 600 alumnae of 90 nationalities.

Is humanity destined to destroy itself, or is it just the wrong people in charge?

A young co-author and friend of mine, determinedly probing human behaviour and its roles in driving ecological overshoot, asked me this morning whether I knew of any work to identify whether humans can theoretically behave in an ecologically 'net positive' way. This young man, a father of young children, had helped me, running a global alliance of organizations and experts to reduce human planetary impact, realize that ecological overshoot is not simply driven by our human numbers (population) and appetites (hyperconsumption)—but more fundamentally by the evolutionary behaviours which had been co-opted for corporate and private gain.

But there are whole global extended networks of people who have always behaved in a relative balance, a net ecologically positive way—actively stewarding and restoring the land and waters. In Spanish they are called 'Los Indígenas.' In North America, where I now live after decades working in Africa, most indigenous people use seven generations thinking: seven generations backward, to learn lessons from ancestors, and forward, to plan to be 'good ancestors' to our children's children's children.

And around the world, despite systemic persecution and often long histories of land theft and genocide, indigenous people protect and restore the lands and waters that they live on and with.

Even in western society, which has tended to celebrate economic and population growth, individualism, competition, resource accumulation and power hierarchies, there are major global networks of people actively working towards communities of people at local, national and global levels to transform to a nature-positive life and work. Most of my own life's initiatives have been focused on this, most recently with the new Global Restoration Collaborative which I started in 2023 to accelerate a regenerative civilization and economy of a billion people by 2050, by bringing together siloed groups in ecosystem restoration, climate restoration, biodiversity conservation and rewilding, youth and women's leadership, and indigenous cultural bridges. And I am integrating this into the Global Evergreening Alliance, a slightly older and dynamic coalition of ecosystem restoration projects, experts and partners working at scale and at speed around the world.

Environmental degradation is not an inevitable outcome of humanity per se. It's only an inevitable outcome of a growthism mindset and global economy of humanity at numbers which crush this planet—3, 4, 5, 6, 7, 8 billion. And nor are greed, procrastination, domination and status-seeking behaviours some intrinsic curse of humanity as a whole. They are intrinsic behavioural traits explicitly captured, incentivized and amplified by systems: by both western rugged-individualism thought, by capitalism, and by resource competition in situations of crowding. There are some bad apples in every crowd. But we also get systems—and leaders—which either encourage or discourage them.

We can change all this. We can choose leaders and systems which learn from the past, learn from each other, plan for the future. The contrasting experiences of Japan and Haiti after earthquakes in 2010-2011 show the power of culture, leaders, and systems in navigating crises.

But it takes significant effort to change the values that so many hundreds of millions of people have been fed all their lives. Much of the affluent western world, with ecological footprints that are choking this planet (Fig. 2), has a perilously short timeframe to undo the damage of these narratives and consciously pursue better paths, with better leaders and better systems.

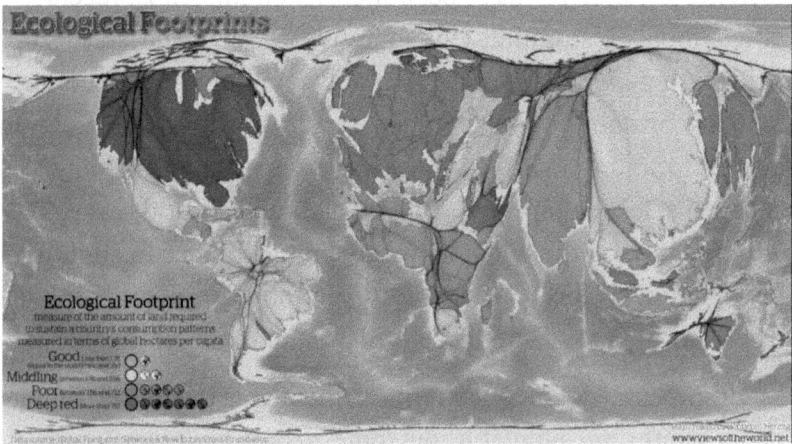

I have long said that democracy, as much as I love it, is not up to the task of navigating the coming decades—unless we can decide now to reform it, reform our economic systems, and accept that our mindsets of comfort, complacency, and cultural entitlement are now defunct and suicidal. And to actively redesign our civilizations and the systems that scaffold them.

Women must lead this call and lead this process. We cannot depend on most of our existing leaders. And, as we all know by now, there is no time to waste.

References (listed in order of use in hyperlinked/ underlined text)

1. https://www.theguardian.com/politics/2023/sep/22/ rishi-sunak-urged-to-stop-attacking-climate-change-committee

2. https://www.theguardian.com/commentisfree/2023/sep/26/ kevin-mccarthy-republican-shutdown-trump

3. https://www.theguardian.com/global-development/2023/ sep/19/climate-action-must-respond-to-extreme-weather-driv- ing-health-crisis-says-who

4. https://www.theguardian.com/us-news/2023/sep/26/ what-government-shutdown-means-2023

5. https://www.theguardian.com/world/2023/sep/22/ canada-wildfires-forests-carbon-emissions

6. https://education.nationalgeographic.org/resource/ key-components-civilization/

7. https://www.abebooks.com/Civilization-Oppression-Wilson-C atherine-ed-Calgary/9007525739/bd

8. https://personal.eur.nl/veenhoven/Pub2010s/2014h-full.pdf

9. https://www.ecologicalcitizen.net/pdfs/epub-086.pdf

10. https://journals.sagepub.com/doi/ epub/10.1177/00368504231201372

11. https://en.wikipedia.org/wiki/Ecological_overshoot

12. https://www.apa.org/monitor/jun04/discontents

13. https://theconversation.com/slow-solutions-to-fast-moving-ecological-crises-wont-work-changing-basic-human-be-haviours-must-come-first-215055

14. https://www.dandebat.dk/eng-klima7.htm

15. https://en.wikipedia.org/wiki/Holocene#:~:text=The%20 Holocene%20corresponds%20with%20the,urban%20 living%20in%20the%20present.

16. https://www.thecollector.com/ first-cities-human-civilization-oldest-cities/

17. https://whc.unesco.org/en/list/364/#:~:text=The%20property%2C%20built%20between%201100,Enclosure%20and%20 the%20Valley%20Ruins.

18. https://www.wgtn.ac.nz/__data/assets/pdf_file/0011/1776422/ Why_do_women_care_more_about_the_environment_than_ men.pdf

19. https://www.scientistswarningeurope.org.uk/discover/ world-scientists-recommend-urgent-actions-for-surviv-al-of-civilization

20. https://academic.oup.com/book/9268/ chapter-abstract/155983746?redirectedFrom=fulltext

21. https://www.smithsonianmag.com/history/ hypatia-ancient-alexandrias-great-female-scholar-10942888/

22. http://physics.ucsc.edu/~drip/7B/hypatia.pdf

23. https://www.smithsonianmag.com/history/ hypatia-ancient-alexandrias-great-female-scholar-10942888/

24. https://thonyc.wordpress.com/2019/01/09/ hypatia-what-do-we-really-know/

25. https://earthobservatory.nasa.gov/features/ OrbitsHistory#:~:text=Galileo%20discovered%20evidence%20to%20support,later%20renamed%20the%20Galilean%20moons).

26. https://www.jstor.org/stable/1086284

27. https://www.laphamsquarterly.org/roundtable/killing-hypatia

28. https://onlinelibrary.wiley.com/doi/10.1111/aec.12391

29. https://journals.sagepub.com/doi/ epub/10.1177/00368504231201372

30. https://en.wikipedia.org/wiki/Ecological_overshoot

31. https://en.wikipedia.org/wiki/Hard_and_soft_science

32. https://www.sciencedirect.com/book/9780126278651/ quantitative-ecology

33. https://thehill.com/changing-america/respect/equality/591070-more-women-in-stem-lead-people-to-label-them-as-soft/

34. https://www.academia.edu/15072411/Potential_impacts_of_ climate_change_on_southern_African_birds_of_fynbos_and_ grassland_biodiversity_hotspots

35. https://onlinelibrary.wiley.com/doi/abs/10.1111/jbi.12714

36. https://www.jstor.org/stable/20751610

37. https://www.bentley.edu/news/ are-men-really-better-suited-success-women

38. https://www.forbes.com/sites/tomaspremuzic/2021/03/07/ if-women-are-better-leaders-then-why-are-they-not-in-charge/?sh=75420e3e6c88

39. https://www.jstor.org/stable/1086284

40. https://www.girlplanet.earth/our-global-voices

41. https://www.wgtn.ac.nz/__data/assets/pdf_file/0011/1776422/Why_do_women_care_more_about_the_environment_than_men.pdf

42. https://hbr.org/2019/06/research-women-score-higher-than-men-in-most-leadership-skills

43. https://www.forbes.com/sites/tomaspremuzic/2021/03/07/if-women-are-better-leaders-then-why-are-they-not-in-charge/?sh=75420e3e6c88

44. https://homewardboundprojects.com.au/about/vision/

45. https://homewardboundprojects.com.au/busara-circle/

46. https://homewardboundprojects.com.au/hb-community/hb1-participants/

47. https://homewardboundprojects.com.au/

48. https://dattnergroup.com.au/fabian-dattner/

49. https://homewardboundprojects.com.au/wp-content/uploads/2023/10/19092023-Media-Release-Women-leaders-unite-to-witness-climate-challenges-up-close_Homeward-Bound_Final-V2-released.pdf

50. https://www.theguardian.com/sustainable-business/2015/jun/29/a-seven-step-guide-to-net-positive

51. https://www.stableplanetalliance.org

52. https://www.haudenosauneeconfederacy.com/values/

53. https://www.romankrznaric.com/good-ancestor/media

54. https://stories.undp.org/10-things-we-all-should-know-about-indigenous-people

55. https://www.stableplanetalliance.org/restoration

56. https://www.evergreening.org

57. https://www.localfutures.org/
    growthism-its-ecological-economic-and-ethical-limits/

58. https://www.visionofhumanity.org/
    contrasting-levels-of-resilience-the-cases-of-haiti-and-japan/

59. http://www.viewsoftheworld.net/wp-content/uploads/2015/11/
    EcologicalFootprintMap.jpg

Prof Phoebe Barnard

Global citizen, Prof Phoebe Barnard, is a global change and biodiversity scientist, policy strategist and filmmaker, intent on accelerating civilizational shift to a future 'where people and the planet actually matter'—where we have re-established a relationship of love for and stewardship of Nature, and of compassion for and tolerance with each other.

Phoebe is campaign advisor to the Global Evergreening Alliance; co-founder and co-leader of the Global Restoration Collaborative; co-founding CEO and advisor, Stable Planet Alliance, founder of the global women's platform on planetary health and personal choice, GirlPlanet. Earth; Affiliate Full Professor of Environmental and Societal Futures and Conservation Biology, University of Washington; Research Associate, African Climate and Development Initiative and FitzPatrick Institute of African Ornithology, University of Cape Town; and film co-producer and science/policy advisor, Transmediavision USA, Inc.

Women's leadership, two-way mentorship and professional development coaching have been core to her work in southern and eastern Africa for 34 years and another seven (so far) back in North America.

Connect with Phoebe on LinkedIn at https://www.linkedin.com/in/phoebe-barnard.

CHAPTER 19
Trailblazing Through Adversity

The Power of Self-Leadership and Personal Branding for Displaced Women Scientists

Rana Mustafa

To my mom, who instilled within me the power of hope, love, and faith. To my kids, Sana and Anas, who give me the strength to thrive through adversity. And to my friend Laial, who always stood by my side and encouraged me to share my story.

Introduction

My journey began in Syria, a land known for its rich history and cultural heritage. I was a university professor with a promising future, surrounded by the warmth of family and the familiarity of home. But life, with its unpredictable twists, had other plans for me. The onset of war, the challenges of displacement, and the battle with breast cancer forced me to redefine my path and reimagine my future.

Prior to the 2011 civil war, Syria boasted renowned scientific, medical, and engineering institutions that were highly regarded in the Arab world. However, this conflict triggered one of the largest mass migrations since World War II. Unfortunately, as of 2022, the Syrian Arab Republic continued to be the primary source of refugees globally. Over 10 years of war has displaced 6.5 million Syrians, comprising over 32% of the country's population.

Amid this backdrop, countless scientists found their careers abruptly interrupted and their expertise at risk of being wasted. In this chapter, "Trailblazing Through Adversity," I provide insights based on my own experiences to demonstrate how personal branding can empower us to not just survive, but also thrive despite life's challenges.

Early Inspirations and Leadership Development

I was always driven by a desire to stand out and be different. In first grade, when my teacher asked me about my dream, I responded that I wanted to be a bride. This answer drew laughter from my classmates, but it was actually rooted in a deeper instinct for self-differentiation. This internal motivation persisted as I grew older, manifesting in unique hobbies and interests. While my peers gravitated towards conventional activities, I found solace in writing poems and short novels and reading history books.

Life was challenging growing up in a rural area surrounded by mountains and olive farms. Each day, I walked 40 minutes across the mountains to reach school—a journey that was especially tough during windy, rainy, or scorching days. My responsibilities outside school were far from typical. As the oldest child and only girl, I had to look after the house and my five brothers while my mother worked on the farm. My mother didn't get the chance to go to school, and her duties inside and outside the house were limitless; she never seemed to rest. This early exposure to the challenges faced by women around me, including the lack of access to education and societal pressures to conform to traditional roles, ignited a determination in me to follow a different path. I discovered that education was my way to change my life and break away from the traditional expectations imposed on women in my community.

Through reading history books, I discovered powerful examples of inspiring Syrian women. Ishtar, also known as Inanna in Sumerian mythology, was a goddess with control over war, love, and fertility. Julia Domna, born around 160 AD in Emesa (present-day Homs), became the wife of Roman Emperor Septimius Severus and the first empress of the Severan dynasty, where she had significant political, social, and philosophical influence. Queen Zenobia of Palmyra, a third-century AD warrior queen, conquered Egypt and Asia Minor, breaking away from Rome.

Learning about these powerful Syrian women and the significant roles they played in ancient societies was both enlightening and disheartening. It was shocking to realize how women's positions had deteriorated over time. Women, who once held power and influence, had become subjugated in a male-dominated society. This stark contrast fueled my determination to pursue education as a tool to change my future.

By the time I reached grade 12, my ultimate goal was to become a physician. However, my family's financial situation posed a significant obstacle. My father told me he couldn't afford to send me to medical school because he also needed to pay for my brothers' education. Faced with this reality, I chose to pursue food engineering instead, not wanting to be the reason my brothers couldn't go to school. That was my first pivot in my career path.

Pursuing food engineering was navigating uncharted waters. Initially, it wasn't a voyage fueled by passion—my sentiments lay dormant, unroused by the lectures and textbooks that filled the first year. With limited options, I faced a decisive moment. The path forward could lead to either failure or success. I made my choice: I would not give up, instead I would explore the unknown territory to find the hidden potential it held.

As I explored the world of food production and food security, my perspective began to shift. The importance of ensuring a stable food supply and the complex interaction between science and agriculture started to resonate with me. I began to value my role as a future food engineer, recognizing the significant impact and necessity of this field.

My academic journey was far from isolated. I was supported by those who had walked this path before me—friends and mentors whose guidance served as a lifeline during challenging times. This fellowship of knowledge-seekers was instrumental in my growth, helping me embrace and excel in my academic career.

My dedication and determination did not go unnoticed. I was honored with multiple awards for academic excellence. This recognition further cemented my reputation within the college, a reputation I was determined to uphold. The crowning achievement was securing a scholarship to pursue a PhD in France, a rare opportunity for women from minority backgrounds like me.

However, pursuing this dream came with significant personal sacrifices. My daughter was just one year old when I began my PhD journey, and her father did not accompany me to France. I faced the heart-wrenching decision to leave my daughter in Syria to be raised by my mother due to personal circumstances. The separation was painful, and the weight of this decision remained with me throughout my time in France. I struggled with separation anxiety and depression, but my love for my daughter ignited my determination to change. This determination to change marked the beginning of my journey into self-awareness and self-leadership, which I later discovered were crucial for navigating adversity.

Driven by the goal of changing my family's future, I was resolved to persevere and finish my PhD. Though the path was difficult, achieving my goal allowed me to enter the academic world as a faculty member and researcher. This phase marked one of the most rewarding and beautiful periods of my life. The satisfaction derived from teaching, the excitement of pursuing my chosen research, and the meaningful relationships I built with my students and colleagues provided me with a profound sense of purpose. Their gratitude served as a powerful affirmation of my impact.

By leveraging my networking skills, academic excellence, and grant writing abilities, I secured two international research opportunities at top universities in Canada. These achievements were especially rare for a university professor from Syria, where academic resources were limited.

From Academic Tranquility to War-Torn Turmoil

The peaceful five years I once led as a university professor in Syria, immersed in the vibrant world of teaching and food engineering, was shattered by the outbreak of civil war in 2011. The stability and fulfillment I had found in my career and personal achievements were suddenly replaced by chaos and the harsh reality of war.

The war was indiscriminate—my home and the university where I had spent years building a career were reduced to ruins. The dreams and achievements I held dear were consumed by the conflict. Amid the rubble, I confronted the painful reality that all that I had worked so hard to build was now lost to the whims of war. Facing this new reality was incredibly challenging, especially after spending over a decade striving to not just advance my own knowledge but to contribute to the wider field of food science.

For five years during the war, I found solace in creativity. When electricity and the internet were unavailable, and necessities were scarce, I took it upon myself to learn guitar and compose haiku poetry. These pursuits weren't just hobbies; they were crucial channels for creativity, providing clarity that sparked new ideas and potential paths forward. Engaging in these creative activities helped me stay mentally strong, enabling me to support my students and community through challenging times.

A New Beginning

When it became clear that the war would not end quickly, I knew I had to make a significant change for my family's future. I dreamed of a new life where my children could grow up without fear and with abundant opportunities. I reached out to universities and funding agencies around the world, hoping to find a way to continue my academic research. Unfortunately, the responses were few—many seemed overwhelmed by the refugee crisis, hesitant to take on more. Despite this, I worked tirelessly, often putting in hundreds of hours to build my online presence and network, determined to improve my professional reputation and achieve my goal. Finally, my efforts paid off, and I received a job offer as a visiting professor in Canada. Moving to Canada meant leaving behind everything familiar, but it was also an opportunity to redefine myself and contribute to the world in new ways.

Embracing Change in Canada

When I came to Canada, I didn't arrive as a stranger; I had already built a strong online reputation as a food scientist and was well-known in the international community. This robust personal brand expedited my job search journey, allowing me to continue my professional pursuits. It also opened doors to numerous opportunities and collaborations, further solidifying my presence and impact in my field. However, arriving in Canada as a single mom with two kids presented significant challenges, including my kids' cultural shock, establishing a personal support network, navigating credential recognition, and adapting to new workplace norms. With each passing day, the aspiration of revitalizing my academic career seemed to drift farther out of reach, leaving me confronted with the daunting prospect of beginning anew. This journey pushed me to my limits but also showed me how to become a stronger and more resilient leader.

In this new chapter, I embraced vulnerability and courage, honing my ability to connect with others empathetically and strategically. These skills didn't just help me adjust to a new career path; they were essential in building collaborative relationships that were key to mutual growth. I transitioned from a role as a researcher overseeing my own projects to one focused on business development and research management for an entire college at the university. This new role facilitated the expansion of my professional network and empowered me to achieve the impactful outcomes I've long strived for.

However, I only enjoyed one year in this role before receiving the devastating diagnosis of breast cancer. This illness forced me to step away from the professional setting for 17 months and ultimately caused me to lose my job. Cancer was a universal signal to start finding my purpose and live every day with excitement as if it was the first day of my life. During that time, I felt compelled to share my story and support other displaced scientists facing similar challenges. This realization led me to found Grow Strong Coaching, aimed at empowering scientists to enhance their personal brands and overcome obstacles in their professional lives.

Reflecting on my journey

My life journey has taught me that leadership is not about titles or circumstances, and it starts by self-leadership. I learnt that navigating adversity requires building four main leadership competencies: Resilience, Endurance, Nurturing, and Acceptance and here are how these competencies help us:

1. Resilience: In life we need mental toughness, adaptability, and the ability to bounce back from adversity. To build resilience, we need to focus on self-reflection to identify moments of success and leverage them. We need to set personal aspirations and measure our progress against our own standards, not others. Such reflection enhances our self-awareness, helping us discover effective strategies for bouncing back. By reviewing past experiences, we can build a personal toolkit that highlights our unique strengths and gifts.

2. Endurance: To handle life challenges, we need to sustain effort and persist over time. Similar to exercising at the gym, building endurance involves gradually pushing ourselves beyond our comfort zones. This continuous stretch develops our capacity to handle longer and more complex challenges over time.

3. Nurturing: It is crucial to enrich ourselves with knowledge, empathy, and self-care, especially during tough times. These practices are crucial for maintaining our well-being and ability to recover from setbacks.

4. Acceptance: When we choose how to navigate challenges, we need to acknowledge and accept the outcomes, even when they are not as we expected. We need to let go of rigid attachments to predetermined outcomes and recognize that the unknown may hold better opportunities and new possibilities.

For career advancement, I discovered the transformative power of personal branding, which led me to my dream job and earned me international recognition as a scientist and leadership coach.

Personal branding for scientists is often misunderstood as self-promotion, but it is truly about defining and communicating our unique identities. It encompasses our values, expertise, and the impact of our research. A strong personal brand enhances credibility, attracts collaborators, and makes our work more accessible to the public. It helps secure funding, attract top talent, and cultivate collaborations with esteemed scientists and institutions. The benefits of personal branding extend beyond networking, significantly contributing to long-term career growth and scientific achievements.

Personal branding is particularly vital for displaced scientists because it helps us overcome unique challenges, such as limited access to professional networks and resources. By showcasing our strengths, resilience, and expertise through a well-crafted personal brand, we become more visible and credible in the scientific community.

Creating a personal brand involves a strategic approach that I have refined through my coaching programs, such as the "8 Steps to Accelerate Your Career with Personal Branding." It starts by defining our unique strengths, values, and goals. The next steps involve sharing our insights and discoveries through various channels such as networking events, conferences, social media, blogs, and public speaking engagements. Creating content on a personal website or a professional profile on platforms like LinkedIn, ResearchGate, or

Google Scholar is invaluable for highlighting research, publications, and outreach activities. Engaging with peers, sharing experiences, and actively participating in conversations within the scientific community also played a crucial role in my personal branding strategy.

By sharing my experiences, I aim to inspire and equip others with the tools they need to thrive, shaping a brighter future for both myself and future generations. Throughout my journey, I've learned the importance of perseverance, resilience, and maintaining a positive mindset. As Rumi said, "You are not a drop in the ocean, you are the ocean in a drop"—a reminder that we each carry within us abundant possibilities and the potential to make a profound impact in this world.

Dr. Rana Mustafa

Dr. Rana Mustafa is a multilingual Scientist, Leadership Coach, Positive Intelligence® Coach and Entrepreneur with over 20 years of expertise in academia, innovation, research and project management. Rana has helped thousands of university students and scientists develop the skills and networks they need to bring their research to market and make an impact beyond the academic world.

In her role as the founder of Grow Strong Coaching and a research facilitator at the University of Saskatchewan, Dr. Mustafa assists researchers with grant writing, business development, and establishing industry partnerships to advance their research initiatives and foster innovation.

Her coaching extends to assisting scientists and sciencepreneurs in building their personal brand, enhancing their professional presence, and improving their self-leadership skills, thereby positioning them as the go-to experts in their respective fields.

Rana's approach to coaching is based in empathy and Positive Intelligence®, providing inclusive, individualized, and authentic coaching that helps people identify and overcome their limitations, develop resilience, and successfully navigate their chosen path.

However, Rana's story extends beyond her professional achievements. As a survivor of war and breast cancer, she brings a unique perspective to her work. She extends her support to professionals grappling with life's challenges, helping them cultivate resilient mindsets, reinvent themselves, and turn adversity into opportunities.

Whether it's through coaching, research or partnerships, Dr. Rana Mustafa is always striving to make a meaningful impact and help shape a better future.

To read about her journey and connect with her on social media, click on this link: https://linktr.ee/growstrongcoaching.

CHAPTER 20

Leading in Uncertainty: As Part of an Organization and as an Entrepreneur

Tanya Maximova

Leadership is to make happen what otherwise would not.

In a world where change is constant and new decisions and pivots need to happen almost every day, the cost of making the wrong move is much lower than the cost of indecisiveness and standing still (which is a decision in itself). Therefore, the ability to make sound decisions, even when you don't have all the information at hand, is becoming the top skill sought after in today's leaders.

In the age of digitization and technological breakthroughs across all industries, uncertainty increases exponentially: we introduce unexplainable AI and quantum computing, where even the laws of physics have duality and are based on probability instead of certitude, new technologies that can disrupt entire industries built in the past century, forcing everyone to operate at lightspeed and adjust as they go.

I was geared up to work in uncertainty, leading digital transformation projects that were so big and complex that it was humanly impossible to determine with certitude every single detail and confirm all the assumptions. As a perfectionist, it took me years to adjust myself as a leader, as a person, to be able to cope with such a level of unknowns. But it also prepared me for the world of entrepreneurship, which would have been impossible for me to take on had I not been ready for it, and I want to share with you the lessons I learned.

Short-term lens versus long-term lens

Starting a multi-year digital transformation project is like performing a brain transplant for an organization. At the outset, questions about business operations are everywhere, but we begin by taking a 10,000-foot view. From this high level, we make educated assumptions, seeing only the start and end points and visualizing the broader path ahead, knowing full well that the specifics will change over time.

As the project unfolds, we dig deeper, testing and confirming these initial assumptions one by one. This process is like gradually sharpening the lens: our "resolution" increases, ambiguity decreases, and a more refined action plan emerges. Every time we adjust, we move closer to a clearer, more precise view.

Through numerous projects across diverse industries, I've grown comfortable with not having all the answers on day one. Instead, I focus on untangling the details progressively, using flexibility and adaptability as essential tools. These skills—learning to pivot through initiatives and daily operations—are something I've honed intentionally, seeing them as vital to successful leadership in dynamic environments.

Why is this important? Every step up the career ladder challenges you to refine your decision-making, pushing you toward a faster, riskier, and more strategic approach based on fewer certainties and more assumptions. The most uncertain position I've ever held was as the founder of my AI healthcare startup, IngrediLens. I was in uncharted waters: a new industry and a groundbreaking technology, faced daily with seemingly impossible decisions. Startups, with their relentless pace, test every skill and bit of knowledge you possess. But it isn't only founders who experience this. Any leadership role involves making tough, high-stakes choices—a challenge especially if you're risk-averse.

If you're not ready to embrace this risk, discomfort might lead you to think, "This isn't for me." But often, that thought is a comforting lie, keeping you safely within your comfort zone. Every period of growth I've experienced started with discomfort: new expectations, unknown territories, and psychological unease—all prerequisites for personal development. Unfortunately, many women are raised to avoid risk, encouraged to be "perfect" rather than "brave."

Reshma Saujani, author of *Brave, Not Perfect*, captured this well and offers this insight in her book:

We're raising our girls to be perfect and our boys
to be brave.

In a world where knowledge doubles every 12 months, we can't afford to wait for perfection. Trying to know everything before acting only delays us, and by then, we may need to start over. True growth, in life and career, means letting go of perfection and stepping into the unknown, ready to act.

When making decisions where the answer is not clear-cut, where no matter which option you choose, there is a level of risk associated with it and allows for vulnerability and trust. When you have that, you have people who are not afraid to make bold moves, to be creative and innovative. So how do you cultivate this in yourself and in your organization, allowing you to operate in a high-paced uncertain world? Looking back at my experience, I see that there are four pillars that create a fertile ground for this culture:

1. Organizational safety

2. Controlled continuous growth

3. Repetitive exposure to uncertainty

4. Compassionate leadership

Organizational safety

I start with this pillar since it is, to my knowledge, the most important one. Truly understanding it can be difficult, but necessary, in order to create a culture of initiative, flexibility, and empowerment. Creativity and innovation, as well as the audacity to take risks and bold actions, can't function in stress. Multiple studies indicate that stress can impair the early stages of the creative process, which often involves brainstorming and ideation. The fight, freeze, or flight response in our bodies generated by stress also creates tunnel vision. This makes it impossible to look at a problem from multiple perspectives and make unusual connections, which is our ability to think creatively.

If we want to come up with ingenious solutions and out-of-the-box thinking, if we want people to feel free to experiment and to be able to voice their ideas without fear, we need to ensure a safe environment that allows them to fail and learn.

During my time as a digital transformation lead, I learned that one of the most important roles of a leader is to be the conductor and the channel of communication, breaking down silos and barriers and enabling their team's initiatives. I would ruthlessly break down the walls between people, departments, and hierarchical structures, giving voice to those who had ideas but weren't heard. This approach allowed me to empower so many individuals and generate so many out-of-the-box innovative ideas with my teams, with incredible buy-in from them. There is no stronger motivation than the intrinsic motivation to bring to completion something you helped create. And when you have the backing of your leader, it gives you courage. In a startup environment, where the pace is so fast you can't always build safeguards, what you can do is create mini experiments to test your assumptions with as little impact as possible, and then move on to bigger ones. A gradual step approach, where you can fail small and test the waters, is possible when there is no other option.

Controlled continuous growth

After you have created a safe environment in your organization, it's time to challenge individuals. As I said before, growth comes out of discomfort, and as a leader, it is my duty to continuously push the people in my teams outside their comfort zones. Every person has infinite potential; the time and mindful practice we spend on specific activities and learning determines our ability to perform better on certain tasks and grow personally and professionally. I am a lifelong learner and believe that we can continue to reinvent ourselves throughout our lives, growing and adjusting to the ever-changing world.

Many of the obstacles we see in front of ourselves are self-fulfilling prophecies; our own perception of our limitations makes us think that we can't possibly be capable or have the necessary skills to embark on the journey we dream about. But the moment we make the first move in that direction, we realize that a journey is comprised of all the steps, and that if we are able to make the first one, and then the second, we are able to reach the end goal. The role of the leader is to create the path, to give a push forward, and to give them strength. It is especially important to give strength to those who, either through past experiences or cultural bias, have been neglected, pushed aside, or overlooked: women, non-binary individuals, people of color, and other visible or non-visible minorities that are left out of the "power circle."

With the people in my teams, I create a roadmap with a phased approach to the activities they will be involved in over time. And with each new phase, they develop a new skillset, gain new responsibilities, and increase their impact and strength little by little. And believe me, it's not a walk in the park; it is a hard journey. But when we look back together after a year or two at the ground they've covered, the personal and professional growth they've had, and the incredible empowerment and impact they feel at the moment, it's definitely worth it.

Repetitive exposure to uncertainty

People naturally avoid ambiguous situations; we are creatures of safety, and uncertainty is the opposite of that. In a startup world, where everything is even more compressed, the sheer number of assumptions and unknowns is mind-boggling. With every decision you make, you don't know right away whether it will bring you to the right outcome; there is no cheat sheet of correct and incorrect answers. Changing that mindset and not expecting to get the perfect score is one of the hardest transformations I had to go through. Usually at work, in an established organization, you have someone who tells you if your decision was a good one: be it your superior, or if you are the CEO, your Board of Directors.

In the incipient stages of starting a company, you are the person you are reporting to; no one will point to a mistake. And if you stop to clarify and analyze everything in detail, you will forever be paralyzed with indecision. You need to learn to identify the core bits of information needed to take a well-enough informed decision and make it. With enough practice, you start spotting the essential pieces of the puzzle, always being able to zoom out and see the big picture and direction. Every time I catch myself over-analyzing and postponing a decision, I go back to the North Star vision, the general direction, and ask which option is more aligned with it. I also ask myself: what is the price of making a mistake at this point, and is the step irreversible? This is a wake-up call, since in business there are very few truly irreversible decisions.

The magic of all this is that the more you get exposed to and forced to take action in uncertain situations, the better and more comfortable you get. I did the same thing with the teams I was leading in my digital transformation projects, which, as you recall in the beginning, have a laundry list of unknowns and assumptions. People who have never before been exposed to such a lack of certitude (usually employees have very clear instructions), after being taken out of their comfort zone with gradual exposure to new skills, new tasks, levels of autonomy and impact, and especially being driven to make decisions that involve a certain level of risk, became more and more comfortable doing it.

Compassionate leadership

One of our fundamental needs as human beings is the need for belonging, of being part of a social structure, and to be accepted by our peers. The longevity research of people living in the five blue zones on Earth, where the population lives longer and healthier than average, discovered the key factors for longevity. There were two common things: waking up every morning with a sense of purpose and being part of a supportive social circle. The community brings us joy and a sense of belonging, which reduces stress levels and ultimately makes us happy. An organization is a mini-society, where employees are the players, and they try to find their place in the group. So, as a leader, creating an environment of support and compassion has long-lasting positive consequences on the team's cohesion, overall morale, and ultimately performance.

The good news is that women are very good at compassion; we are socially and naturally conditioned for it. The bad news is that nobody really understands its importance yet. So, it's like a dormant superpower that we use without actually acknowledging how crucial it is for building a successful team, having creative ideas and a safe space, and building a positively driven organizational culture based on purpose and collaboration. My hope is that soon this will become one of the sought-after skills assessed during interviews for leadership positions.

Compassion is about truly connecting with the other person, being there 100% in the conversation. When you talk to someone, listen and don't answer emails and messages at the same time. You will be surprised by the result – the person feels truly heard, supported, and understood.

Compassion is also about letting go of your ego, your ideas, and giving space to the team. As a leader, you can come up with ideas on how a challenge can be overcome. However, from personal experience, instead of pushing your view on others, ask your team, have a conversation with them and, again, listen. I can't stress enough the importance of listening. If you present your team with a solution right away, they will, of course, accept it, but it will limit their own creativity and motivation, or even the courage to come up with new suggestions. By giving them space to contribute, you co-create, and this is a beautiful process that fosters a motivated and engaged group and results in unforeseen and ingenious solutions, which you would never be able to devise by yourself.

Conclusion

I've had to reinvent myself a few times throughout my life, and every single time it was a journey of self-discovery, of failure and success, of pushing the boundaries of what I thought was possible, and of rediscovering my potential. Sir Edmund Hillary said, "It is not the mountains we conquer, but ourselves." And it's so true.

With an ever-increasing pace of new discoveries, new technologies, and new information overall, the search for permanence and certainty leads to fear and disappointment. The desire to control the things we can't predict is doomed to defeat, so we better learn to adapt and adjust. And we do it by enabling a safe space where people can take risks, learn, experiment, take on more responsibilities, and be able to make bold decisions, by fostering continuous growth and compassionate leadership.

Tanya Maximova

Tanya Maximova is an Engineer and MBA with a passion to solve complex problems, ambition to have an impact and commitment to bring ideas to life. With over 15 years of global experience as Senior Delivery Lead in Digital transformation with a demonstrated history of working in a broad selection of manufacturing, trade, service industries and financial institutions as well as cross-industry business transformations. Leading Projects and Programs ranging from Six sigma quality improvement to ERP implementations, streamlining the Digital landscape of businesses and increasing their profitability and efficiency. Through her career, she has developed expertise in large scale Business & Technological transformation, Operational improvement, Change management, Organizational and Business Optimization.

Continuously searching for innovative technical solutions to everyday problems, she founded an AI Health Tech startup called IngrediLens. It is an AI enabled mobile platform designed for individuals with food allergies, intolerances, and dietary restrictions, enabling them to discover new products and enjoy food with confidence. This was an opportunity to build something that makes a difference and empowers people who live with food restrictions and seems like the purpose-driven initiative she is always striving for. The role of a founder required her to go through several mindset shifts, has put to a test all her skills and knowledge, but also created a green field opportunity to start something of impact she truly cared about.

Connect with Tanya at https://www.ingredilens.com.

CHAPTER 21

The End of the "Power Dead-Even Rule"

Victoria Vojnovich

I dedicate this chapter to women leaders who are committed to building a legacy of leadership and who realize that gender equity is not pie. There are enough slices for everyone.

I remember my first day as a people leader. I was to lead a service delivery team for a popular networking product with a large, global install base. My reporting manager called a meeting of my department, invited all of the team (easier back then as we were all in the office every day), and even brought the donuts and coffee. I was a little nervous, even though I had been mentored and given plenty opportunities to lead while not having direct responsibility for a team. I had led projects, teams, presented to senior leadership, and met with a Senior VP every month for 12 months in an executive mentoring program. So I was well prepared, and it was time to formally step into the role.

I entered into the department meeting with my manager, and we grabbed a cup of coffee and sat down. He then introduced me to the team and announced that I would be the new department manager. I put on my biggest smile, thanked him for the opportunity, told my new team how happy I was to have the role, and how excited I was to get to know the team and show what we can do for the business. Everyone shook my hand and welcomed me, and I thought to myself, "First hurdle overcome. This will be a great day."

Then I walked back to my office to start the role. On my desk was a yellow Post-it note with a handwritten message. It quoted a passage from the Bible, "But I suffer not a woman to teach, nor to usurp authority over the man, but to be in silence," followed by a request to be transferred out of the department immediately. So much for the first hurdle overcome. I closed my office door and took a deep breath. This man did not know me, and in fact had only met me for the first time only 30 minutes prior. And he already wanted out. And what was worse, he was the most senior engineer and viewed as the leader of the team! Was he speaking only for himself, or was this going to be the first of a stream of Post-it notes? Every member of this team was male. Were they even going to give me a chance?

Remembering my mentoring about addressing issues before they grow into monsters, I spent about 15 minutes thinking what I wanted to say to him, and then walked down the hall and asked him for a few minutes of his time. I thanked him for letting me know that he was uncomfortable working for a woman manager and acknowledged his desire to leave the department. I then told him that I would speak to HR to inform them of his request, and that I would work with him to expedite that process as soon as he found a position he wanted to transfer to. In the interim, I informed him that my expectation was that he would do his job with his usual zeal, and that he would not inspire a revolt by the rest of the team. He told me that he expected to find a new position within a month. He ended up working on my team until his retirement, almost five years later. Several years later, this man, who could not imagine working for a woman leader, died unexpectedly, and I went to his funeral to pay my respects. His wife recognized me, and when I gave her my condolences in the receiving line, she said to me, "I don't think he ever told you, but you were his favorite manager throughout his entire career." I was both floored and honored.

I think about this story whenever I see statistics that show that despite more women graduating with STEM degrees and increases in recruiting efforts for women in STEM, the percentage of women in STEM careers still hovers in the 33-35% range. And most recent statistics show that women in STEM leadership roles is about 25%. While some of this is directly related to the "boys club" thinking that men will promote men and that men do not want to work for women, there is another culprit – *The Power Dead-Even Rule.*

In the early 1980s, Dr. Pat Heim coined the term, "The Power Dead-Even Rule." Dr. Heim researched behaviors of men and women in the workplace. She noticed that men would tear each other apart in a meeting and then offer to share a beer after work. Women's behavior was completely different. Women who were raised to "play nice" focused on having all women appear to be equal in skills, abilities, and levels, and when a woman was promoted, she was ostracized and rejected. Also, when women were given leadership roles, they took on their male counterparts' traits and behaviors rather than bringing their own strengths such as empathy and emotional intelligence to their role. Fearful that the number of opportunities for women in leadership were limited, women tended not to promote and advance other women.

Early in my leadership career, I attended a woman leader in tech conference where I got to hear Dr. Heim talk about this concept. At that very moment, I made the commitment that I would not be this type of leader. I wanted my leadership legacy to be a leader who grows more leaders. And I wanted many of them to be women. Upon returning to my company, I joined the newly formed women's employee resource group. It was the best decision of my career. In the years that followed, I have been a member, a leader, or sponsored as an executive in the women's employee resource group at every company I have worked for. If we want to change the landscape of leadership in STEM for women, we have to work together as women and with women.

Employee Resource Groups (ERGs) are those safe spaces where women can support each other, connect with each other, and grow skills. In many STEM companies, even today, a woman may be the only woman in her department, and even extended team. They may not even know the other women in the organization, and most certainly in the company if it is a large company. I encourage women to join their ERG as soon as they join the company. Many early in career employees are reluctant to join ERGs because they are focused on impressing their leadership and learning the ways of working at their companies. As a leader, I encourage every new employee to join an ERG and get to know others outside their teams. For women in my organizations, I bring them with me to the meetings!

Because the structure of an ERG is employee led and volunteer run, the organization gives women who may not have the opportunity in their teams a safe space to grow leadership skills, get mentored, and get recognized by other senior leaders in the company. Every opportunity that I have been given in my career after that first manager job was the result of a connection made through an employee resource group. I have found executive sponsors, life-long mentors, and colleagues who have become friends long after we stopped working together. In their most recent "Women in the Workplace" report, McKinsey and Company noted that ERGs are a necessary tool for recovering women's careers that were stalled during the Covid-19 pandemic. If you don't have a women's ERG at your workplace, step up and start one. I'll be happy to partner with and support you!

Another way that we break the *Power Dead-Even Rule* is to build personal boards of directors. Employees have used mentors for decades, but there is real value in having a board of directors rather than one mentor only. First, each of us needs to be both developed and challenged. One mentor does not have time to do both. Second, a board of directors brings you diversity: diversity of thought, diversity of experience, diversity of situation, and diversity of support. Finally, as the CEO of You, Inc., you have many needs. More mentors are better than one!

When creating your board, think of the seats you want to offer up and what those directors bring to the table. My board of directors contains the following seats (including how I select my directors):

1. Sponsorship and Influence
   a. Who at a senior level supports you?
   b. Who is influential?
   c. Who offers guidance?
   d. Who connects you to resources?

2. Purpose and Motivation
   a. Who inspires you with fresh ideas?
   b. Who motivates you to make a difference?
   c. Who validates your work?
   d. Who role models taking action?

3. Knowledge and Expertise
    a. Who builds your level of knowledge?
    b. Who is an expert in the areas you want to develop?
    c. Who shares best practices or innovations?
    d. Who broadens your best perspectives?

4. Personal Development
    a. Who makes you a better you?
    b. Who challenges you?
    c. Who gives you candid feedback?
    d. Who tells you things you don't want to hear?
    e. Who pushes you to be better?

5. Personal Support
    a. Who encourages you?
    b. Who listens while you vent?
    c. Who gets you back on track when needed?
    d. Who can you be yourself with?

6. Balance
    a. Who encourages your health and mental well-being?
    b. Who helps you to contribute to your community?
    c. Who helps you grow and learn as a person?

When building your personal board of directors, keep the following things in mind:

1. Keep your board limited to six people. I gave suggestions here on roles, but you can create roles that are meaningful for you. However, a board that is larger than six will be hard to manage; you will be overwhelmed with too many inputs.

2. Schedule regular meeting times with your directors, but not all at once (unless it's a really big decision). Respect the time that

your directors give you; let them know what you want to talk about and what you are looking for from them.

3. Pay it back and pay it forward! Be a woman leader who grows more women leaders.

4. Board members will change based on their availability and your needs. Just remember to keep your board diverse.

You have many sources of potential board members. Consider people from your current company, previous company(s), leaders in your community, your faith community, civic organizations, etc. You want your directors to be people you respect, people who will be direct and honest with you, and people who are committed to your growth.

As women leaders in STEM, we have to focus on breaking the *Power Dead-Even Rule* and continue to build a legacy of great women leaders. Employee Resource Groups and Personal Boards of Directors will help us achieve this. Justice Ruth Bader Ginsberg was once asked, "'When will there be enough [women on the Supreme Court]?" Her answer was, "When there are nine" Some people were shocked with her reply and she then countered with, "But there'd been nine men, and nobody's ever raised a question about that." (RBG. Sep 21, 2020) While we will always need to compete with the Boys Club mentality and need male allies to grow the ranks of women leaders in tech, as women we have to remember that leadership spots are not like slices of pie, where only one slice per pie is allocated to a woman. Let's take RBG's thinking and fill as many leadership slots as we can with women that we can sponsor, mentor, and cheer for their success.

Victoria Vojnovich

Victoria Vojnovich is the founder and CEO of iEmpowHer, LLC. She is committed to bridging the global gender equality gap for women in the workforce and re-imagining the future of work to create equitable workspaces for all. Having spent her entire career in Tech, Victoria understands the value of mentorship and sponsorship for women, who still only make up 23% of the tech workforce (and unfortunately, only 10% of leadership roles in tech). In the five companies where she worked, she has led or sponsored the women's employee resource groups to support women's professional growth and development in a safe and nurturing space.

Victoria also leads a global cloud services transformation office for an Indian headquartered global systems integrator (GSI). She holds a Professional Certified Coach (PCC) designation from the International Coach Federation, among multiple certifications and designations. She is also a member of Chief, the largest network of senior women executives. Vojnovich has a B.S. degree in Mathematics and Computer Science from Penn State University, an M.A. degree in Women's and Gender Studies from North Carolina State University and is completing her PhD program at North Carolina State University, where her dissertation research topic is "Re-Imagining the Future of Work for Mid-to-Late Career Women."

Connect with Victoria at www.iempowher.net.

Afterword

Your Chapter Awaits

Dear Reader,

As you turn this final page, you might find yourself reflecting on the journey you've just experienced through these words. Each story you read is a tapestry of dreams, struggles, triumphs, and the relentless spirit of its creator. Now, imagine a world where your story joins these ranks – where your voice, your experiences, and your unique perspective are shared and celebrated.

This is not just an invitation; it's a call to action from the curator of this book, Cathy Derksen, the owner of <u>Inspired Tenacity</u>. Cathy believes in the power of stories to transform, inspire, and connect humankind. More importantly, she believes in your story and its potential to make a significant impact on the planet.

Why wait for "someday" to tell your story? The time is now, and the world is ready to listen. Whether it's a tale of adventure, a deeply personal memoir, a groundbreaking idea, or a story that has been quietly growing in your heart, it deserves to be told.

Inspired Tenacity specializes in turning visions into reality. Cathy understands the journey of transforming a personal narrative into a published book – it's a journey of courage, creativity, and breaking through fears. Cathy and her team are dedicated to guiding you through every step of this exhilarating process, from the initial draft to the moment your book is held in the hands of eager readers across the globe.

Join her vibrant community of authors, a diverse group of storytellers who have dared to make their voices heard. You'll discover a supportive network of mentors, editors, and fellow authors who are all committed to the success of your story.

Take the leap. Embrace the thrill of seeing your own story in a book. Contact Cathy at Inspired Tenacity, and embark on this remarkable journey together. Your story matters, and the time to share it with the world is now.

p.s. Remember, every great story begins with a simple decision to start writing. Yours is no different. Let's make it happen, together.

www.ingramcontent.com/pod-product-compliance
Lightning Source LLC
Chambersburg PA
CBHW070924210326
41520CB00021B/6792